Prospects in diagnosis and treatment of breast cancer

PROSPECTS IN DIAGNOSIS AND TREATMENT OF BREAST CANCER

Proceedings of the Joint International Symposium on Prospects in Diagnosis and Treatment of Breast Cancer, 10–11 November 1993, Munich, Germany

Editors:

Manfred Schmitt
Department of Gynecology, Technische Universität München, Klinikum rechts der Isar, Munich, Germany

Henner Graeff
Department of Gynecology, Technische Universität München, Klinikum rechts der Isar, Munich, Germany

Günther Kindermann
Department of Gynecology, I. Frauenklinik der Ludwig-Maximilians-Universität München, Munich, Germany

Co-editors:

Fritz Jänicke
Department of Gynecology, Technische Universität München, Klinikum rechts der Isar, Munich, Germany

Thomas Genz
Department of Gynecology, I. Frauenklinik der Ludwig-Maximilians-Universität München, Munich, Germany

Björn Lampe
Department of Gynecology, I. Frauenklinik der Ludwig-Maximilians-Universität München, Munich, Germany

 1994

Excerpta Medica, Amsterdam – London – New York – Tokyo

International Congress Series No. 1050
ISBN 0 444 81707 7

This book is printed on acid-free paper.

Published by:
Elsevier Science B.V.
P.O. Box 211
1000 AE Amsterdam
The Netherlands

Library of Congress Cataloging in Publication Data:

Joint International Symposium on Prospects in Diagnosis and Treatment
    of Breast Cancer (1993 : Munich, Germany)
        Prospects in diagnosis and treatment of breast cancer :
    proceedings of the Joint International Symposium on Prospects in
    Diagnosis and Treatment of Breast Cancer, 10-11 November 1993,
    Munich, Germany / editors, Manfred Schmitt, Henner Graeff, Günther
    Kindermann ; co-editors, Fritz Jänicke, Thomas Genz, Björn Lampe.
            p. cm. -- (International congress series ; no. 1050)
        Includes index.
        ISBN 0-444-81707-7 (alk. paper)
        1. Breast--Cancer--Congresses. I. Schmitt, Manfred, 1947-
    II. Graeff, Henner. III. Kindermann, Günther. IV. Title.
    V. Series.
        [DNLM: 1. Breast Neoplasms--diagnosis--congresses. 2. Breast
    Neoplasms--therapy--congresses. W3 EX89 no. 1050 1994 / WP 870 J74
    1994p]
    RC280.B8J65 1993
    616.99'449--dc20
    DNLM/DLC
    for Library of Congress   94-13767
                                                                      CIP

In order to ensure rapid publication this volume was prepared using a method of electronic text processing known as Optical Character Recognition (OCR). Scientific accuracy and consistency of style were handled by the author. Time did not allow for the usual extensive editing process of the Publisher.

Printed in the Netherlands

Preface

Breast cancer strikes one out of every 8—10 women in industrialized countries. At the time of primary treatment the majority of patients already has occult micrometastatic disease due to the capacity of tumor cells for invasion and early hematogenic spread. Therefore every second breast cancer patient will develop metastases and eventually succumb to the disease. There are still problems in identifying the patients at risk for relapse, and even among those with progressive disease, the course of disease may be extremely variable.

When we planned our meeting on *"Prospects in Diagnosis and Treatment of Breast Cancer"* we had three goals in mind: 1) to bring together research-oriented clinicians and tumor biologists; 2) to focus on mechanisms or factors of tumor biology with prognostic relevance; and 3) to evaluate new aspects of treatment oriented at tumor biology. We thus tried to follow the need for more specific, less invasive and less toxic methods of treatment.

The demand for new prognostic factors serving as selection criteria in reaching treatment decisions has increased, since it was shown recently that adjuvant therapy even leads to higher cure rates in breast cancer patients. The prognostic impact of a given factor to indicate early relapse and death may also be indicative for its respective tumor biologic weight, thus giving a clue for the elaboration of therapeutic strategies oriented at tumor biology. Indeed, the first aspects of a molecular therapy oriented at the interference with certain factors of tumor biology were presented at the meeting. In this context it may well be remembered that the classical anti-hormonal treatment of breast cancer — as transferred into clinical practice years ago by William C. McGuire and Heinrich Maass among others — is still an outstanding example of a specific therapy of minor toxicity and invasiveness.

Impressive structure-function analyses of tumor suppressor genes, of proteases and of adhesion proteins on the DNA/RNA and the protein level were presented at the meeting. These and other studies yielded essential insights in the cellular transformation process and the regulation of malignant behavior of the tumor cell. Examples for the application of a *"Synthetic Biology"* using antisense oligo-nucleotides and synthetic peptides were also given and represent promises for the interference with the invasiveness and the proliferation rate of breast cancer cells.

In March 1983, the first international symposium was held in Munich under the auspices of the Wilhelm Vaillant Foundation and was chaired by Josef Zander. That symposium was entitled *"Early Breast Cancer"* and its main emphasis lay on breast-conserving surgery. The editors of the proceedings (Early Breast Cancer, Histophathology, Diagnosis and Treatment. J. Zander and J. Baltzer eds., Berlin, Heidelberg, New York, Tokyo: Springer Verlag, 1984) state in their preface that breast-conserving surgery must be regarded as a significant breakthrough and must no longer be overlooked. The symposium and the ensuing book paved the way for further endeavors at conservative treatment of breast cancer.

The editors wish to express their sincere gratitude to the Wilhelm Vaillant Foundation and to the Chairman of the Managing Committee, Ministerialdirigent Alexander Freiherr von Hornstein. It was again the Foundation's generous support that made it possible for the symposium to take place and for this book to be published. We also wish to thank the Bayerische Akademie der Wissenschaften and Syndika Monika Stoermer for making it possible for the meeting to take place at this historical site.

Invited speakers came from all over Europe and the United States. However, the backbone of the distinguished list of speakers was furnished by the members of the Concerted Action *Clinical Relevance of Proteases in Tumor Invasion and Metastasis* within the Biomed-1-Program of the Commission of the European Communities. The organizers want to thank cordially all speakers, chairmen, discussants and guests for their efforts to reach the aforementioned goals of the meeting.

The Editors
Henner Graeff
Günther Kindermann
Manfred Schmitt

Memorial to Prof. Dr. Ing., Dr. med. Wilhelm Vaillant

On January 28, 1993 Professor Dr. Wilhelm Vaillant died at the age of 83. He was a colourful personality — impulsive, full of dynamic vitality, a man of mark. The Wilhelm Vaillant Foundation, which owes its existence to his unselfish generosity, organized this celebration in his honour and remembrance.

Wilhelm Vaillant's life can be divided into five major stages:
— childhood and youth in Chemnitz and Dresden (1909—1928);
— the studies in Dresden (1928—1934);
— the German Railway official (1934—1945);
— the entrepreneur (1945—1966); and finally,
— the medical doctor and patron (1966—1993).

This short enumeration of the various periods of his life already illustrates that Wilhelm Vaillant was an exceptional man; a restless man who had achieved a lot but who was never fully content with what he had achieved; a man who was constantly seeking new challenges and mastering them; a man full of life and — just like life — full of contradictions.

He was generous and petty at the same time. When colour television was introduced, he gave each of his employees a colour TV set for Christmas and he even took them to their homes himself. As a purchaser, he would bargain about the price of nails or screws. He could be most charming and engaging — if he felt inclined. But it was better not get into a fight with him, although this could happen quite easily — to be unpunctual or careless at work was reason enough. If things were going too slowly he would become active himself.

Working with Wilhelm Vaillant was definitely not easy. However, the sometimes ruthless ways he used to realize his own ideas were possibly a major reason for his success. Another factor was certainly that Vaillant was able to think in context; he was flexible and could foresee events. He was a skilful but trying partner in negotiations. He preferred to negotiate near the border of failure.

Wilhelm Vaillant was a craftsman and technician, and he understood his job. He immediately grasped complex design plans for contracts he had awarded. His ideas for possible changes were always to the point, precise and clear. Technicians working for him were sometimes enthusiastic about the swiftness with which Vaillant saw technological problems and was able to solve them — with technical knowledge and a lot of intuition. With his business profits he immediately started new ideas, always on the solid base of his technical know-how, always dauntless but with a never-failing sense for the possible and an infallible feeling for the point of profitable retreat. At an age when other people think of retirement he fulfilled a dream he had had since his youth — he studied medicine.

Prof. Dr. Ing., Dr. med. Wilhelm Karl Theodor Vaillant (1909—1993)

It all began on April 22, 1909 in Chemnitz when Wilhelm Karl Theodor Vaillant, as was his full name, was born. His father was General Technical Manager of an electric plant at the state railroad, his mother a housewife. A fellow high school student remembers him as a plain and shy boy — well liked but inconspicuous and by no means remarkable. Wilhelm Vaillant often described his childhood in Chemnitz as a time of privation. In fact, some recollections suggest narrow and modest circumstances in his life. From early on he had to contribute to the livelihood of his family. He tended pigs, bred rabbits and had to do all kinds of other chores. Thus he felt, at an early age, the burden of responsibility on his young shoulders.

He completed his schooling by graduating from the natural science branch of the Realgymnasium-Annaschule in Dresden. According to accounts and pictures Wilhelm Vaillant was then a rather serious, reserved, perhaps even a sad and melancholy person. From 1928 to 1934 Wilhelm Vaillant studied electrical engineering in Dresden. Everything went according to plan, and he completed his studies in 1934 with a degree as a certified engineer. Even then it was apparent that he was a "technician with a tendency toward the Fine Arts", which is how Wolf Posselt from the German TV station ZDF saw him many years later. Wilhelm Vaillant was active in the student council where he was in charge of cultural affairs. He organized visits to the theatre and dance evenings. He published articles regularly in the newspaper "Sächsische Volkszeitung" on technical, scientific and cultural events.

It was the theatre that interested him but even more so the opera. When looking back, does the fact that Wilhelm Vaillant worked as a scene-shifter in the theatre and as a waiter in the restaurant at the railroad station to finance his studies not appear like the starting point of a biographical continuous thread? From the small world of the German railroad that was familiar to him from childhood, to the larger world of the stage he was now scrutinizing, and when he shifted to TV decades later the bell of his entrepreneurial began to toll.

But this time had not yet come. So far Vaillant had not veered from the track. He did practical professional training to become a senior civil servant in engineering with the "Reichsbahn", the German railroad. In May 1936 he passed the examination to be an engine driver, a title he was especially proud of his entire life. After passing the 2nd state examination, certified engineer Wilhelm Vaillant became — in the autumn of 1937 — senior civil servant in the engineering department of the Reichsbahn. He then learned a lot: planning, surveying, building — activities he profited from later on. During World War II Wilhelm Vaillant was made head of the light-technical department of the electro-technical institute in Munich. There he led a group commissioned to develop blackout measures. With a thesis in his special field, the extension of blackout measures to the field of infrared radiation, he was awarded a doctor's degree at the technical university in Karlsruhe under Prof. Dr. Rudolf Georg Weigel. These studies brought Vaillant into contact with a field of technological research which was to play an important role in his later life.

After the war, Vaillant's experience and his diverse talents went together — just like a puzzle — to give the picture of a strong entrepreneurial personality. Together with the lawyer Dr. Hans Ritter he set up a light-bulb company in Munich and founded the RIVA-Copier-Werk GmbH. The cinema boom of the years after the war then showed results, an upward trend was visible. However, all of a sudden turnover was decreasing, imperceptibly at first but then more increasingly evident. At this time Wilhelm Vaillant sensed the dawning of television. In the late autumn of 1958 construction of the TV studios in Unterföhrung near Munich began. In 1961 the "Bayrische Rundfunk", Bavaria's radio station, became the first long-term tenant. Shortly afterwards Vaillant managed to make a well-timed move: The RIVA-TV studios were sold to the Bavarian radio station. At the same time, part of it was again rented in order for the Bavarian studio of the German TV station ZDF to move in.

In the shortest time possible, right next to the old area which was now property of the Bavarian radio station, Wilhelm Vaillant constructed studios which offered the most modern production technology. The introduction of a modular scene-shifting system and the drastic reduction of operating distances enabled continuous production known only at the BBC. In 1966, after having ventured a production of his own, Vaillant sold the new studios to the ZDF.

Wilhelm Vaillant then turned to a new part of his life. After having registered in the winter term of 1961/62 for medicine at the Ludwig-Maximilians-Universität in Munich he then finally started his studies seriously. With the verve that was so typical for him he took up this new challenge. At the same time he increasingly enjoyed spending his great financial means to the advantage of the general public. In 1969, at the age of 60, Vaillant finished his medical studies with very good results. In 1970, he received his doctor's degree under Prof. Dr. Georg Maurer at the Medical Faculty of the Technical University Munich with a thesis on "Experiments on early diagnosis of breast cancer by thermography". Now he had found the topic which was to rivet him for the rest of his life — the fight against breast cancer in women. Here he did real pioneer's work, especially in the field of early diagnosis by the employment of additional technology. After having received his license to practise in 1970, he became an honorary professor for biomedical technology in 1972.

During his studies, he had already given valuable help to the medical schools of Munich universities, especially by donating sophisticated technical equipment. In 1981 he set up the foundation that is named after him which intends to promote medical research as well as the building and expansion of institutions for preventive health care. He built up an institute for preventive medical care himself which was later incorporated into the Technical University Munich due to its excellent scientific standard. At the First Department of Obstetrics and Gynecology of the Ludwig-Maximilian-University in Munich he established the Wilhelm-Vaillant-Unit for early diagnosis of breast diseases and donated it to this University Hospital in 1985.

Wilhelm Vaillant has made a vast contribution to the society and the country. The Bavarian Distinguished Service Order and the Great Distinguished Service Cross of the Distinguished Service Order of Germany, which were bestowed on him in 1969 and 1985, respectively, are visible signs of his merits.

In the Wilhelm Vaillant-Unit and in the Wilhelm Vaillant-Foundation his name and this part of his life's work will live on. In our hearts we will keep him in grateful reminiscence as a colourful, headstrong, dynamically vital, impulsive and striking personality with an unmistakable profile of his own.

Alexander Freiherr von Hornstein

Chairman of the Managing Committee
of the Wilhelm-Vaillant-Foundation

Contents

xiv

Prognostic factors, new aspects of treatment

Tumor cell proliferation, growth and differentiation

3

Can oncogene and tumor suppressor genes aid in assessing node-negative mammary tumors?

Heinz Höfler[1,2]*, Michael J. Atkinson[1] and Arne Luz[1]

[1]GSF-Institute of Pathology, Neuherberg, D-85758 Oberschleissheim; and [2]Institute of Pathology, Technical University of Munich, School of Medicine, Ismaninger Straße 22, D-81675 München, Germany

Abstract. Reliable prognostic parameters describing node-negative mammary cancer are a prerequisite for diagnosis and therapy. Much hope has been raised by recent advances in the molecular biological understanding of cancerogenesis. However, the practical benefits of oncogene and tumor suppressor gene analysis have yet to be realized. The diagnostic potential of assays of genetic alteration in mammary cancer is reviewed here. A number of the genetic changes characterized show potential; but their use as prognostic parameters requires careful validation using standardized methodologies. At this time caution is urged in the use of these parameters as prognostic indicators.

Introduction

There are extensive overlaps in the risk of recurrence of both node-negative and node-positive breast cancer [1,2]. In a substantial minority of patients with lymph node-negative breast cancer the course of the disease is subsequently seen to be highly aggressive. Consequently it is imperative to identify patients with low stage lymph node classification (N0) who may prove to be at risk from metastatic disease. More accurate discriminatory parameters are required for this to be realized. This is a particularly urgent clinical problem, brought about by the increased rate of detection of in situ and small invasive cancers (<2.5 cm) made possible by improved imaging technologies. As a direct consequence of the increased detection efficiency there is a rising demand for prognostic parameters with which to allow these women and their clinicians to make appropriate therapeutic choices [2,3].

With the advent of molecular biological techniques numerous alterations in the genome of tumor cells have been identified, and subsequently accorded a high priority in the list of potential disease markers. However, the rate of appearance of new candidates has greatly exceeded that of clinical studies proving or disproving their applicability. Consequently many of the proposed markers are untested, and of those that have been adequately tested little or no advantage over classical methods has been demonstrated [4,5]. However, if sufficient effort is devoted to the painstaking phase of testing, some of the markers offering promise may subsequently

---

*Address for correspondence: Heinz Höfler, Institute of Pathology, Technical University of Munich, School of Medicine, Ismaninger Strasse 22, D-81675 München, Germany. Tel.: +89-4140-4160. Fax: +89-41805-165.

find their way into the routine diagnostic laboratory. We have attempted to review and condense the current status of a number of potential markers. In doing so, we have focussed upon their applicability to mammary tumors, and have intentionally not discussed the validity of individual markers in other tumor types.

Genetic alterations in tumor cells: mutation of oncogenes and tumor suppressor genes

Oncogenes are dominantly active genes, defined by their ability to transform immortalized cells in vitro [6,7]. Tumor suppressor genes, sometimes incorrectly termed antioncogenes [8], are recessively acting genes which undergo loss of function mutation. The wild-type allele retains the capacity to suppress the transformed phenotype in vitro. The most relevant tumor suppressor genes have been identified as germline mutations, tightly linked to dominant familial cancer syndrome phenotypes [8–12]. Oncogenes and tumor suppressor genes correspond to the somatically mutated alleles of normal cellular genes. In sporadic tumors the most frequently encountered somatic gene alterations affecting oncogenes are amplifications (often, but not necessarily, resulting in overexpression) or mutations (primarily point mutations). Point mutations, combined with loss of heterozygosity (LOH) at the gene locus, are the primary cause of loss of function in the case of the recessively acting tumor suppressor genes (see reviews cited above). The characteristic LOH at a tumor suppressor gene locus has been used effectively in sporadic tumors to identify future suppressor loci [13]. It is important to note that hitherto unidentified tumor suppressor genes involved in the etiology of mammary cancer may reside in the various LOHs described in breast cancer [5].

Int-2

The oncogene int-2 [14] was originally identified as a site of mouse mammary tumor virus (MMTV) integration in the genome of transformed murine mammary cells. The gene product is a member of the fibroblast growth factor family and appears to function primarily in regulating epithelial growth. Amplifications affecting the int-2 locus were found in 11% of breast cancer cases (node-negative cases 30/79) by Henry et al. [14]. During follow-up (7 to 60 months, mean 24 months) all 12 patients with int-2 amplification (but only 20% of patients without amplification) died. The prognostic significance for node-negative patients also seems to be significant, according to former studies [14]. Since in the study of Henry et al., fewer than 1.8% of node-negative patients contained amplified int-2, the predictive value relates to only a very small number of patients. Overexpression of the int-2 gene product may not itself be involved in tumorigenesis as int-2 is infrequently expressed in breast cancer. The locus may merely serve as a marker for the amplification of other genes located within the chromosome 11q13 amplicon [5], including the closely linked bcl-1/ PRAD-1/Cyclin D1 oncogene (see below).

Receptor for insulin-like growth factor 1

The IGF-1-receptor is a common cellular component, essential for the regulation of proliferation. Expression of the receptor has been demonstrated in over 90% of primary breast cancers tested. The gene is infrequently affected in mammary tumors, with an increased copy number being observed in only 2% of 975 cases (node status not reported) studied by Berns et al. [15]. Eight patients with tumors having a very high copy number ($\geq$20 copies) had a shorter median overall survival (42 months; range, 14—120+) than 4 patients with tumors having a low amplified (3—10 copies) IGF-1-R gene copy number (median, 77 months; range, 19.5—98+).

HER-2/neu/c-erbB-2

The neu oncogene (also termed HER-2 or c-erbB-2) is a transforming oncogene member of the epidermal growth factor receptor family [5]. Initial studies suggested that amplification/overexpression of the HER-2/neu gene could furnish important clinical information on mammary tumor status. However, a review of the available literature by Clark and McGuire in 1991 [16] revealed no correlation between HER-2/neu status and disease-free time or overall survival. Immunohistochemical analysis of expression continues to provide equivocal results, although follow-up criteria in these studies are usually more comprehensive. In the 1992 study of Schroeter et al. disease-free survival was shown to be longer in patients (nearly 50% of whom were node-positive) with little or no HER-2/neu expression (74%) compared to those with overexpression (44%) [17]. However, this difference was only evident within the first 2 years of follow-up. In a study of 230 node-negative cases [18] HER-2/neu immunohistochemistry had no prognostic value. A more recent study by Press et al. [19], using node-negative patients stratified according to T-classification, observed a higher odds-ratio of prognosis for HER-2/neu overexpression in T1A (13.3) as compared to T2A (1.9). The authors stress the necessity of immunohistochemistry with reliable reagents and recommend the future usage of computerized image analysis for a standardized scoring of the result.

ras

The *ras* family of genes encode GTPases involved in cytoplasmic signal transduction. Oncogenic activation of *ras* is caused by point mutations at a limited series of codons [20,21]. This highly selective mutational spectrum has made *ras* the object of many studies, most of which have relied upon inexact technologies to screen large patient samples. Although observed at high frequency in many other tumor types, such as adenocarcinoma of the colon and lung, *ras* gene mutations are rarely encountered in human breast cancer [5]. Here we must again urge caution in the interpretation of this essentially negative finding. Current state-of-the art DNA analysis has not been performed on significant numbers of samples.

c-myc

The nuclear transcription factor c-*myc* is in part responsible for determining the rate of apoptotic cell death. Overexpression of c-*myc* due to translocation or amplification is believed to be responsible for the acquisition of an oncogenic function [22,23]. Analysis of c-*myc* expression at the protein or mRNA levels, is complicated by their short half-lives [5]. A study of 95 node-negative patients by Berns et al. showed strong data indicating that amplification of c-*myc* was a significant predictor of reduced relapse-free survival [24]. The 5-year actuarial relapse-free survival fraction was 0.69 ± 0.05 for normal c-*myc* gene copy number (84 cases) and 0.18 ± 0.12 for amplified c-*myc* gene (11 cases). In a similarly sized sample containing 48 node-positive and 30 node-negative patients amplification of c-*myc*, Henry et al. [14] demonstrated that c-*myc* amplification had no prognostic significance.

Bcl-1/PRAD1/cyclin D1

The bcl-1/PRAD1 locus at chromosome 11q13 encodes a protein product cyclin D1 closely involved in regulating cell-cycle progression [25]. The locus is already well documented in mammary cancer, being in close proximity to the int-2 locus. In a prospective study overexpression of cyclin D1 mRNA was detected in 45% of 124 breast cancer specimens from patients undergoing simple or subcutaneous mastectomy [25]. This promising area is the subject of ongoing study.

Retinoblastoma gene RB1

The prototypic tumor suppressor gene RB1 (located on chromosome 13q14) participates in the regulation of the cell cycle [26,27]. Although breast cancer does not appear to be a relevant risk in survivors of retinoblastoma [28], circumstantial evidence suggests that RB1 could be involved in mammary cancer. Allele loss of chromosome 13, consistent with inactivation of tumor suppressor loculi, is found in approximately 25% of breast carcinomas. However, to date no association between RB1 loss and clinical outcome has been tested.

p53

The tumor suppressor gene p53 (located on chromosome 17p13) produces a nuclear protein necessary for ensuring fidelity of DNA replication and partition to progeny cells [29,30]. For review see W. Deppert in the following chapter. Mutational inactivations with LOH have been recorded in almost all tumor types studied, and loss of p53 may be a prerequisite for malignant transformation. In mammary tumors a region of chromosome 17p spanning the p53 gene is deleted in over 50% of informative cases of breast cancer [5]. Recent immunohistochemical analysis of p53 for the prognosis of node-negative breast cancer gives cause for cautious optimism [31–33]. Three recent studies have indicated a significant association between p53 antigen expression and reduced survival. Allred et al. [34], using frozen sections, found that 13% of their 452 patients with T 1—3cm, demonstrating p53 immunoreactivity. The median 5-year survival of this group was 69% compared to 91% for those with immunohistochemistry-negative tumors. Silvestrini et al. [35] studied 256 subjects, of whom 44% were positive. The 6-year survival of the p53 positive patients was 69% vs 91%. In an additional study Friedrichs et al. [36] conducted a 60 month follow-up. In a subpopulation of his patients, 54 (mostly T>1) showed a significantly lower survival that was associated with a positive p53 immunohistochemical reaction. However, in two further studies no significant predictive value could be found. In the study of Lipponen et al. [37] follow-up data over at least 10 years were used to retrospectively study 83 subjects, and Andersen et al. [38] followed 99 patients for a mean of 46 months. From the published data it cannot be decided whether inclusion in the studies of different proportions of T1 tumors could be responsible for the discordant results.

On the whole caution is advised in interpreting the available data, and it cannot be concluded that there is a strong correlation between positive p53 immunohistochemistry and p53 gene mutation in breast cancer [39].

Future markers

Cell adhesion molecules

As Hedrick et al. [40] discussed recently, proteins mediating cell adhesion may suppress tumorigenesis through their influence upon cell growth, differentiation and/or invasion. Three immunohistochemical studies indicate the complete loss of the E-Cadherin adhesion molecule in breast cancer. The loss of expression is almost exclusively limited to the infiltrating lobular type tumor [41,42,43]. This finding correlates with the recent observation of a high frequency of E-Cadherin mutation in diffuse gastric cancer [44]. The prognostic value of such a phenomenon has yet to be studied.

nm23

The nm23 gene has been implicated as both an invasion suppressor and as a tumor suppressor. The gene encodes a transcription factor capable of binding to regulatory regions of c-*myc* [45]. Reduction of nm23 expression seems to occur early in breast cancer [46]. Clinical correlates are lacking at this time.

HNPCC

The gene responsible for maintaining some aspects of DNA proofreading has recently been shown to be mutated in hereditary nonpolyposis colon cancer. The molecular consequences of this mutation appears to be loss of control over microsattelite stability throughout the genome. This generalized genomic instability has been found in colorectal carcinoma, gastric carcinoma and endometrial carcinoma. In breast cancer there was no evidence for mutations of the HNPCC gene [47], but mutations of other genes fulfilling a similar role should not be discounted.

BRCA1

Restriction fragment polymorphisms have been used to demonstrate linkage between the "BRCA1" locus on chromosome 17q21 and breast and/or ovarian cancer [48]. Risk estimates for breast cancer (49% by age of 50 years, 90% by age of 70 years), may indicate a mendelian inheritance in affected families. It has been estimated that 1 in 200 women carry the BRCA1 allele which imparts 85% chance of contracting the disease [49].

Conclusion

According to the present state of art there is — except for the BRCA1 allele as a familial marker — no simple high risk marker in sporadic breast cancer. This relates especially to the node-negative patients with low T stage. Probably more carefully designed and stratified studies may show more useful results in the future. This relates also to methodological problems (specificity, qualification) of the assays. To date studies of oncogenes and tumor suppressor genes during follow-up and related to the therapy applied are lacking. In conclusion, the clinical use of oncogenes and tumor suppressor genes as markers for prognosis and individual therapy cannot be recommended yet.

However, even if those methodological problems of studying single "markers" might be overcome, a rapid success might not be expected for principal reasons. The molecular biological approach to pathology of neoplasia is not a simple search for markers. It is a completely new approach to the analysis of pathology in general. Processes until today only described by morphological terms will be replaced by molecular events. Considering as one example the meaning of cell adhesion molecules in understanding the growth behavior of neoplasia demonstrates the complexity at this new level of research: even in this cellular limited field there is a huge network of interplaying molecules [50] which has to be studied carefully in basic research of cell biology. There is no rapid clinical success by the molecular pathological approach.

References

1. Miller WR. Br J Cancer 1992;66:775—776.
2. Dorr FA. Cancer 1993;71:2163—2168.
3. Swain SM. N Engl J Med 1993;328:1635—1634.
4. Walker RA, Varley JM. Cancer Surv 1993;16:31—57.

5. Van de Vijver MJ. Adv Cancer Res 1993;61:25–56.
6. Bishop JM. Cell 1991;64:235–248.
7. Hunter T. Cell 1991;64:249–270.
8. Knudson AG. Proc Natl Acad Sci USA 1993;90:10914–10921.
9. Marshall CJ. Cell 1991;64:313–326.
10. Weinberg RA. Science 1991;254:1138–1146.
11. Bryant PJ. Trends Cell Biol 1993;3:31–35.
12. Levine AJ. Annu Rev Biochem 1993;62:623–651.
13. Lasko D, Cavenee W, Nordenskjöld M. Annu Rev Genet 1991;25:281–314.
14. Henry JA, Hennessy C, Levett DL, Lennard TWJ, Westley BR, May FEB. Int J Cancer 1993;53: 774–780.
15. Berns EMJJ, Klijn JGM, van Staveren IL, Portengen H, Foekens JA. Cancer Res 1992;52:1036–1039.
16. Clark GM, McGuire WL. Cancer Res 1991;51:944–948.
17. Schroeter CA, De Potter CR, Rathsmann K, Willighagen RGJ, Greep JC. Br J Cancer 1992;66:724–728.
18. Bianchi S, Paglierani M, Zampi G, Cardona G, Cataliotti L, Bonardi R, Ciatto S. Br J Cancer 1993;67:625–629.
19. Press MF, Pike MC, Chazin VR, Hung G, Udove JA, Markowicz M, Danyluk J, Godolphin W, Sliwkowski M, Akita R, Paterson MC, Slamon DJ. Cancer Res 1993;53:4960–4970.
20. Egan SE, Weinberg RA. Nature 1993;365:781–783.
21. Lowry DR, Willumsen BM. Annu Rev Biochem 1993;62:851–891.
22. DePinho RA, Schreiber–Argus N, Alt FW. Adv Cancer Res 1991;57:1–46.
23. Meichle A, Philipp A, Eilers M. Biochim Biophys Acta 1992;1114:129–146.
24. Berns EMJJ, Klijn JGM, van Putten WLJ, van Staveren IL, Portengen H, Foekens JA. Cancer Res 1992;52:1107–1113.
25. Buckley MF, Sweeney KJE, Hamilton JA, Sini RL, Manning DL, Nicholson RI, deFazio A, Watts CKW, Musgrove EA, Sutherland RL. Oncogene 1993;8:2127–2133.
26. Goodrich DW, Lee W-H. Biochim Biophys Acta 1993;1155:43–61.
27. Hollingsworth RE, Hensey CE, Lee W-H. Curr Opin Genet & Devel 1993;3:55–62.
28. Eng C, Li FP, Abramson DH, Ellsworth RM, Wong FL, Goldman MB, Seddon J, Tarbell N, Boice JD. J Natl Cancer Inst 1993;85:1121–1128.
29. Donehower LA, Bradley A. Biochim Biophys Acta 1993;1155:181–205.
30. Lane DP. Nature 1993;362:786–787.
31. Yandell DW, Thor AD. Diagn Mol Pathol 1993;2:1–3.
32. Thor AD, Yandell DW. J Natl Cancer Inst 1993;85:176–177.
33. Jensen RA, Page DL. Human Pathology 1993;24:455–456.
34. Allred DC, Clark GM, Elledge R, Fuqua SAW, Brown RW, Chamness GC, Osborne CK, McGuire WL. J Natl Cancer Inst 1993;85:200–206.
35. Silvestrini R, Benini E, Daidone MG, Veneroni S, Boracchi P, Cappelletti V, Di Fronzo G, Veronesi U. J Natl Cancer Inst 1993;85:965–970.
36. Friedrichs K, Gluba S, Eidtmann H, Jonat W. Cancer 1993;72:3641–3647.
37. Lipponen P, Ji H, Aaltomaa S, Syrjänen S, Syrjänen K. Int J Cancer 1993;55:51–56.
38. Andersen TI, Holm R, Nesland JM, Heimdal KR, Ottestad L, Børresen A-L. Br J Cancer 1993;68: 540–548.
39. Lohmann D, Ruhri C, Schmitt M, Graeff H, Höfler H. Diagn Mol Pathol 1993;2:36–41.
40. Hedrick L, Cho KR, Vogelstein B. Trends Cell Biol 1993;3:36–39.
41. Rasbridge SA, Gillett CE, Sampson SA, Walsh FS, Millis RR. J Pathol 1993;169:245–250.
42. Gamallo C, Palacios J, Suarez A, Pizarro A, Navarro P, Quintanilla M, Cano A. Am J Pathol 1993; 142:987–993.
43. Moll R, Mitze M, Frixen UW, Birchmeier W. Am J Pathol 1993;143:1731–1742.
44. Becker K-F, Atkinson MJ, Reich U, Nekarda H, Siewert JR, Höfler H. Human Molecular Genetics

10

1993;2:803−804.

45. Marx J. Science 1993;261:428−429.
46. Steeg PS, De La Rosa A, Flatow U, MacDonald NJ, Benedict M, Leone A. Breast Cancer Res Treat 1993;25:175−187.
47. Peltomäki P, Lothe RA, Aaltonen LA, Pylkkänen L, Nyström-Lahti M, Seruca R, David L, Holm R, Ryberg D, Haugen A, Brøgger A, Børresen A-L, de la Chapelle A. Cancer Res 1993;53:5853−5855.
48. King M-C, Rowell S, Love SM. JAMA 1993;269:1975−1980.
49. Editorial. Nature Genet 1993;5:101−102.
50. Kemler R. Trends Genet 1993;9:317−321.

Functional analysis of tumor suppressor p53

Wolfgang Deppert

Heinrich-Pette-Institut für Experimentelle Virologie und Immunologie an der Universität Hamburg, Martinistraße 52, D-20251 Hamburg, Germany

Introduction

"Tumor suppressor genes", "anti-oncogenes", or "recessive oncogenes" are terms used to describe genes which can restrain tumor growth. Each of these terms describes a different aspect of the negative control of tumor growth. Hence these terms are not freely interchangeable. A common denominator, however, is that elimination of their function is required for tumor growth. One can assume that multistep carcinogenesis requires both the action of activated oncogenes as well as the inactivation of tumor suppressor genes. The prototype of a tumor suppressor gene is the retinoblastoma gene, which seems to be involved in negative regulation of cell growth by modulating the activity of the transcription factor E2F [1]. The list of tumor suppressor genes is rapidly growing, and comprises genes involved in diverse functions, ranging from transcription to cellular adhesion [2,3]. p53 has been added to this list not too long ago. Meanwhile, p53 has been recognized as probably being one of the most important genes in the development of human cancer, as mutations in the p53 gene reflect the most common genetic alteration in human cancer cells [4,5].

p53: from tumor antigen to oncogene

p53 was discovered in 1979 as a cellular protein associating with the transforming protein of the small DNA tumor virus simian virus 40 (SV40), the SV40 tumor antigen (T-Ag) [6]. Later on, p53 was also found in many other nonvirally transformed tumor cells, but not in normal cells, and hence was termed a cellular tumor antigen. Improved methods of detection later on demonstrated that p53 was also present in normal cells, albeit in grossly reduced levels. The function of p53, both in normal and in tumor cells, remained an enigma for quite some time [7]. Around 1984, several laboratories demonstrated that the expression of p53 in normal cells is regulated in a cell cycle dependent manner, suggesting a role for p53 in cellular proliferation. The view of p53 as a regulatory protein in the cell cycle was strongly supported by experiments performed by Mercer and colleagues [8], who showed that expression of p53 is required for the transit of resting cells from G_0 to G_1.

At around that same time, a number of groups provided evidence for p53 being an oncogene: transfection of p53 into primary cells led to immortalization of these cells, while cotransfection of p53 with an activated *ras* oncogene resulted in conversion of the cells to a fully transformed phenotype. Furthermore, transfection of p53 expression vectors into a variety of cells indicated that overexpressed p53 induced effects related to cell transformation or tumorigenesis: in normal Rat 1 fibroblasts, p53 conferred a tumorigenic phenotype onto these cells; p53 enhanced the tumorigenic phenotype of a weakly tumorigenic Abelson murine leukemia virus transformed cell, and it increased the metastatic potential of murine bladder carcinoma cells [9].

p53: from oncogene to tumor suppressor

Later it was discovered that the p53 expression plasmids, used in the studies described above, all had encoded mutant p53 [10]. Repetition of these experiments with vectors encoding wild-type p53 then generated completely different results, leading to an inverse interpretation of the role of p53 in cellular transformation and tumorigenesis: wild-type p53 not only failed to immortalize primary cells, or to transform these cells in cooperation with *ras*, wild-type p53 was even able to suppress these events when cotransfected together with mutant p53, *myc*, or the adenovirus E1A gene in immortalization or transformation assays [11,12]. Furthermore, transfection of wild-type p53 into p53-negative tumor cells led to a reversion of the tumorigenic phenotype and to a growth arrest of these cells at G_1 in the cell cycle [13]. Using temperature-sensitive mutants of p53, or wild-type p53 expressed under an inducible promoter, it was confirmed that indeed overexpression of wild-type p53 in tumor cells caused a reversible growth arrest, suggesting that wild-type p53 may have a negative effect on cell proliferation [14]. These experiments, together with observations described below, established wild-type p53 as a repressor of cell transformation and tumorigenesis.

The classification of p53 as a tumor suppressor alleviated a long standing problem in interpreting the discrepancy between the postulated oncogenic functions of p53 and the observation that in Friend virus induced erythroleukemia in mice an appreciable number of tumor cell clones had totally lost the ability to express p53, very often through severe rearrangements in the p53 gene. Those Friend cells still expressing p53, in virtually all cases expressed a mutant p53 [15–17]. The simplest interpretation of these data now was that loss of a wild-type p53 function was an important step in the generation of Friend erythroleukemias.

The studies revealing the tumor suppressor functions of p53 in the rodent system greatly aided the studies of the role of p53 in human tumors, as it had been found that many human tumors carried mutations in the p53 gene. The p53 alterations observed ranged from total lack of p53 expression in some tumors to overexpression of a mutated p53 in many others. Expression of a mutated p53 very often was accompanied by the loss of heterozygosity (LOH) [18,19]. LOH can be considered a property typical for tumor suppressor genes, indicating the elimination of a wild-type p53 function by mutational inactivation of one p53 allele coupled with the loss of the other allele.

However, one has to keep in mind that mutant p53 can have an oncogenic function of its own, i.e., a mutation in p53 might create a truly oncogenic p53, which simply has not lost a tumor suppressor function but actively contributes to the process of malignant transformation. In fact, the concept of a "gain of function" for p53-mutations is demonstrated by two already mentioned properties of mutant p53: induction of a tumorigenic phenotype in a weakly tumorigenic Abelson murine leukemia virus transformed cell line by transfected mutant p53 clearly demonstrated an oncogenic property of this mutant p53, since these cells did not express any endogenous p53. In addition, the increase of the metastatic potential of murine bladder carcinoma cells after transfection of mutant p53 strongly argued for a dominant-positive role of mutant p53 in tumorigenesis.

The apparently high percentage of p53 mutations, leading to "gain of function" mutant p53 proteins, might also provide an answer to the intriguing question of why p53 mutations are so common in the development of cancer: most p53 mutations are point mutations, i.e., single hits. As all p53 mutants analyzed so far have lost the p53 "suppressor function", an additional "gain of function" implies that a single point mutation would score as "two hits for one". Although this is an attractive hypothesis, the high frequency of p53 mutations might also be explained by assuming that alterations in the p53 gene represent a very prominent restriction point ("bottle neck") in the development of a tumor.

p53 mutations in human cancers

The human p53 gene, located on chromosome 17p, has been analyzed in a wide variety of primary tumors, xenografts and cell lines derived from tumors. So far, p53 mutations have been found in virtually all cancer types looked at, with the interesting exception of a few neural tumors [4,5]. p53 mutations in human tumors have a number of characteristic features: firstly, most of them are missense point mutations giving rise to an altered protein. This implies that such an altered p53 had been selected during development of the tumor. Secondly, mutations in p53 are not randomly scattered over the protein, but cluster in certain areas of the protein. p53 contains five domains, which are evolutionary highly conserved among all species analyzed so far [20]. The majority of p53 mutations are clustered between amino acids 130 and 290 (out of 393) (Fig. 1). Within this area, most mutations are confined to regions II–V out of the five conserved regions of the p53 protein [4,5]. The fact that mutations cluster within cross-species conserved domains firstly suggests that they eliminate or change important functional domains of the p53 protein. However, the finding that different conserved domains are hit already suggests that mutations at different sites of the p53 gene alter different functions of p53. The third important aspect regarding p53 mutations is that there are at least three mutational "hot-spots", affecting codons 175, 248 and 273 [19]. The distribution of these "hot-spots" is quite

Figure 1

p53 Landmarks

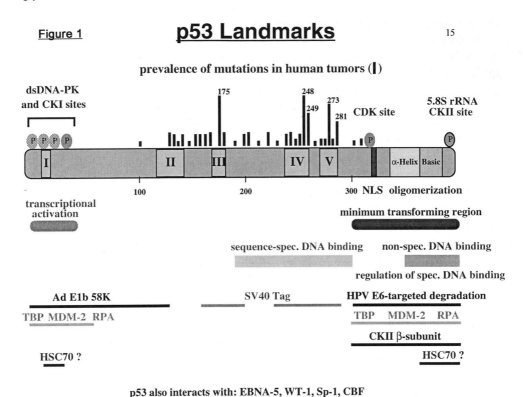

Fig. 1. Structural domains of human p53. Roman numerals represent the five regions of p53 that are conserved from all vertebrates. The main nuclear localization signal is shown (NLS), known phosphorylation sites are indicated above (P). The squares in the center indicate residues mutated in human tumors (hot spots are identified by amino acid number). Shown below is the current information concerning various domains of p53 for biological activities, p53-DNA interactions, and p53-protein complex formation (Ad E1b 58K is the adenovirus E1b p58 protein; TBP is the TATA-box binding protein; MDM-2 is the mouse double minute protein 2; RPA is the eurkaryotic replication protein A; CDII β-subunit is the respective subunit of casein kinase II; HSC70 refers to the heat shock cognate protein; EBNA-5 is the Epstein-Barr virus nuclear antigen 5 protein; WT-1 is the Wilms Tumor protein 1; Sp-1 is the respective transcription factor; CBF is the CAAT-box binding factor).

different among different types of cancer. For example, mutation of codon 175 is not seen in lung tumors, although it is quite common in many other tumors, especially in colon carcinoma [5]. This is remarkable and could either reflect that different mutagens or other environmental aspects are involved in the development of these tumors. Alternatively, different selective pressures for promoting cell growth or outgrowth of a tumorigenic cell might favor different mutational events in different tissues. The most striking example of specificity of p53 mutations is found in hepatocellular carcinoma (HCC) induced by nutritional exposure to aflatoxin, possibly coupled with infection by hepatitis B virus (HBV) in Southern China and parts of Southern Africa. About 50% of HCC in these areas exhibited p53 mutation at the third base pair position of codon 249 (most of them G to T transversions), leading to substitution of Arg by Ser in the mutant p53. In contrast, codon 249 mutations are rare in HCC from patients having a normal risk for their disease [6,21,22]. The importance of p53 mutations for the development of tumors is underscored by the analysis of p53 in inherited forms of cancer such as the Li-Fraumeni syndrome. Li-Fraumeni syndrome affected families have a high incidence of cancer and exhibit germ-line mutations in one p53 allele. Occurrence of cancer coincides with the loss of the other wild-type p53 allele [6,21,22].

Biochemical and functional properties of wild-type and mutant p53

Comparative analyses of wild-type p53 and several mutant p53 proteins both in the rodent and in the human system have helped to build up a list of properties distinguishing wild-type and mutant p53 (Table 1). The most important one, and the one used in many diagnostic analyses, results from the fact that many, but by far not all, mutant p53 proteins display an altered conformation which can be recognized by monoclonal antibodies specific for p53 in a mutant conformation [23]. Perhaps due to such a conformational alteration, wild-type and mutant p53 often differ in their interaction with cellular or viral target proteins. For example, whereas wild-type p53 forms a strong complex with SV40 large T, mutant p53 does not. Conversely, mutant p53 often strongly binds to the cellular 70-kDa heat shock proteins (70-kDa hsc), a property not observed with wild-type p53 [24]. It is clear that these parameters can only provide a gross distinction between the wild-type and the mutant phenotype of p53, and that functional parameters are needed for further classification. This area of p53 research is just picking up momentum. The frantic search for biochemical activities of p53 established that p53 is a DNA and RNA binding protein. The

Table 1. General properties of wild-type and mutant p53[1]

Property analyzed	Wild-type	Mutant p53
Reactivity with monoclonal antibody		
PAb 1801	+	+
PAb 1620	+	-
PAb 240	-	+
Binding to SV40		
Large T antigen	+	-
Binding to cellular		
70-kDa heat shock protein (70-kDa hsc)	-	+
Nonsequence-specific DNA binding	+	±
Sequence-specific DNA binding	+	-
Binding to MAR/SAR DNA elements	-	+
Transactivation	+	-
Half life	~20 min	>3 h

[1]Human p53.

binding of p53 to nucleic acids is complex: the C-terminal end of p53 encodes a domain which mediates the nonsequence specific interaction of p53 with ds- and with ss-DNA, as well as with RNA [25–27]. These properties are common to both wild-type and mutant p53, although mutant p53 seems to be somewhat impaired in ds-DNA binding [26,28]. The C-terminal end of p53 also seems to be involved in the RNA and DNA reannealing activity ascribed to wild-type p53 [27]. In terms of biological function, the notion that wild-type p53 displays characteristics of a transactivator protein (with a transactivating domain localized to the very N-terminus of p53) has drawn considerable attention, and intensified the search for DNA elements sequence-specifically recognized by p53. Several such sequence elements have been identified. Although most of them more or less match a consensus DNA sequence established by El-Deiry et al. [29], there is at least one sequence element specifically recognized by wild-type p53 which hardly fits the consensus [30]. Thus it is difficult to imagine how p53 can specifically bind to all these different DNA sequences. The mystery of sequence-specific DNA binding by p53 is becoming even greater considering that the DNA binding domain of p53 now has been localized to the central portion of the p53 molecule, which does not contain a homology to any of the known DNA binding motifs of DNA binding proteins [31,32]. Furthermore, sequence-specific DNA binding of p53 seems to be "cryptic", and requires "activation" via phosphorylation of a C-terminal casein kinase II (CKII) phosphorylation site, or interaction of the p53 C-terminus with a monoclonal antibody (PAb421), probably mimicking the interaction of p53 with a, so far unknown, cellular factor [31]. Despite all these problems, there now is little doubt that p53 functions as a transactivator in the cell, and several putative target genes have been identified (see below). Again, it is important that the transactivator function is largely confined to wild-type p53, i.e., most mutant p53 proteins have lost this function.

p53 as guardian of the genome

The most important question in the analysis of p53 is the definition of its function in a normal cell. Several lines of evidence have indicated that p53 is an important regulatory element in normal cellular proliferation and development [33]. This view got shattered by the finding that mice made deficient for p53 expression (p53 null mice, p53 knock-out mice) develop normally [34]. These mice, however, are prone to an early occurrence of a variety of tumors. This suggested that p53 was dispensable for normal cell growth and for development, but somehow was required for the control of the cells' genetic stability. At around the same time, an observation made already in 1984 by Maltzman et al. was rediscovered: irradiation of cells expressing a wild-type p53 induces metabolic stabilization of p53, leading to its accumulation, and to G_1 arrest of the irradiated cells [35]. This observation then was extended to other genotoxic agents, establishing p53 as a cell cycle check point protein in G_1 in response to DNA damage [36–40]. Suddenly, everything fell into place. The current view of p53 now is that it acts as a "guardian of the genome", preventing progression of cells through the cell cycle, when DNA damage has occurred, thus allowing for DNA repair. Several independent lines of evidence have established that this watch-guard function is dependent on a functional wild-type p53, and is lacking in p53 deficient cells, e.g., in cells established from p53 null mice [36–40]. Less clear is the mechanism of the p53 induced G_1 growth arrest. There is evidence for induction of growth arresting genes, like the gadd45 gene, mediated by the p53 intrinsic transactivator function [39]. However, a direct interference of p53 with cellular DNA synthesis still remains an alternative option, as negative regulation of viral and cellular DNA synthesis by p53 has been described [39]. Very recently, the binding and functional elimination of replication protein A (RPA) has been reported from several groups [41]. RPA is a multisubunit complex, and binding of RPA to single-stranded DNA may be the initial step in DNA replication, required for unwinding of DNA. Thus, by binding and functionally eliminating RPA, p53 might be able to block initiation of cellular DNA synthesis.

p53 in normal cell growth and development

Although the role and function of p53 as a guardian of the genome at the moment has attracted enormous attention, there is evidence that this is not the only function of p53, implying that p53 is not dispensable for normal cell growth and development. Early experiments already demonstrated an involvement of p53 in normal cell cycle progression. Microinjection of anti-p53 monoclonal antibodies into serum-starved resting cells, stimulated to proliferation by readdition of serum, prevented entry of the cells from G_0 into G_1 [8,42]. Similarly, progression of cells, growth-arrested in the restriction point of G_1 by isoleucine-starvation was blocked by microinjection of anti-p53 monoclonal antibodies after addition of full medium [43]. Thus these experiments strongly argue for a requirement of p53 for cell cycle progression. This view is further supported by experiments of Shohat et al. [44], who demonstrated that expression of p53 anti-sense RNA led to the cessation of DNA synthesis and finally to cell death. An additional argument for p53 being required for cellular proliferation might be deduced from the observation that p53 levels in growth-arrested cells are very low, but rise significantly upon entry of the cells into the cell cycle and reach a maximum in S-phase [45].

Several models have been proposed to reconcile the apparently contradictory roles of p53 in cell cycle regulation. According to the conformational hypothesis put forward by Milner [46], p53 can assume a growth-inhibitory (G_0) and a growth-promoting (G_1) conformation. At the molecular level, it has been suggested that p53 might interact with growth-promoting as well as with growth-inhibitory targets, and that the net outcome of these interactions is dependent on the level of p53, with relatively low levels of wild-type p53 furthering a growth-promoting, and high levels inducing a growth-inhibitory response [47]. As p53 is a sequence-specific trans-activator protein (see above), a concentration dependent induction of different cellular targets might be envisioned.

Recently, it has been proposed that p53 and a cellular target of p53, mouse double minute 2 (mdm2), are functionally coupled by an autoregulatory feedback loop [48]. The mdm2 gene is a cellular oncogene, whose product, the MDM2 protein, was found to bind to p53 and to abolish wild-type p53 transactivator function [49]. The elimination of at least some wild-type p53 function(s) by MDM2 is also suggested by the finding that the mdm2 gene is amplified in certain human sarcomas [50], and that overexpressed mdm2 allows the growth of cells expressing moderately elevated levels of wild-type p53 [47,51]. Regulation of mdm2 expression seems to be under the direct control of p53, as the first intron of the mdm2 gene contains a p53 responsive element consisting of two imperfect p53 consensus binding sites [48,52]. Accordingly, elevated expression of wild-type p53 leads to upregulation of mdm2 expression both in vitro and in vivo [47,48,52]. As MDM2 binds p53, it then is assumed that this binding abolishes wild-type p53 transactivation, thereby connecting p53 and mdm2 in an autoregulatory feedback loop. Such a mechanism could explain that p53 levels can rise during cellular proliferation without inducing a growth arrest. If such a mechanism existed, one would predict a coordinated expression of mdm2 and p53 during cellular proliferation, resulting in the binding of significant amounts of p53 by the MDM2 protein. This, however, has not yet been demonstrated unequivocally.

The p53 nonproducer pre-B cell line L12, has provided an interesting system to analyze the postulated role of p53 in differentiation. In these cells, p53 expression has been abolished by integration of Abelson murine leukemia proviral DNA into the p53 gene [53]. These cells are only weakly tumorigenic in syngenic mice insofar as initially formed tumors regress within a relatively short time. Introduction of a wild-type p53 gene into L12 cells induced terminal differentiation, as measured by the production of cytoplasmic μ protein in vitro, and IgG immunoglobulin in vivo. Consequently, these cells no longer were tumorigenic at all [53]. The regressor phenotype of the parental L12 cells most likely is due to the fact that these cells get induced to differentiate in vivo, as measured by μ-gene rearrangement (without μ protein synthesis) when grown as tumors, but not when the cells are grown in vitro. Thus the reintroduced wild-type p53 pushes this differentiation process one step further. Interestingly, introduction of mutant p53 into these cells strongly enhanced their tumorigenic phenotype, thus providing evidence for the dominant-oncogenic activity of mutant p53, as already outlined [53]. As mutant p53 expressing L12 cells in in vivo tumors did not show any μ-gene rearrangement, one can assume that the dominant oncogenic activity of mutant p53 in these cells results from a differentiation arrest.

How the activities of wild-type p53 in proliferation and in differentiation of normal cells fit into the picture obtained from the analysis of p53 null mice, namely that p53 is dispensable for these processes, is not yet explicable. However, elimination of p53 expression in a stem cell might still allow the use of alternate pathways, which are no longer available for differentiated cells.

Activities of mutant p53

The dominant-oncogenic effect of mutant p53 in L12 cells is not an exception. Indeed, it now is assumed that "gain of function" mutations in p53 comprise a large fraction of all p53 mutations [10,54]. So far, mutant p53 proteins mostly have been scored for losses of activities compared to wild-type p53, and its dominant oncogenic function was mostly explained by trans-dominance, i.e., the elimination of p53 wild-type function by competition of mutant p53 (which usually is present in grossly enhanced levels as compared to wild-type p53) for cellular targets, or by formation of heterologous oligomers between wild-type and mutant p53, in which the wild-type p53 adopts the mutant conformation [55]. Clearly, these mechanisms are not applicable in cells which are devoid of any wild-type p53 expression, like the L12 cells. Thus dominant-oncogenic mutant p53 proteins must exert activities on their own. Very little, so far, is known about such activities of mutant p53.

Like wild-type p53, mutant p53 associates with nuclear structures. As mutant p53 still interacts with nuclear structures (i.e., the cellular chromatin and the nuclear matrix) [26], and binds to DNA in a nonsequence-specific manner, we analyzed mutant p53 for alternate interactions with DNA, defined both by sequence and structure of the DNA. Using a target-bound DNA binding assay developed in our laboratory, we demonstrated that mutant p53, but not wild-type p53, exhibited specific and complex DNA binding to DNA fragments displaying characteristics of nuclear Matrix Attachment Region (Scaffold Attachment Region) DNA (MAR/SAR elements) [56]. These DNAs, generally 0.3 to several kb in lengths, anchor cellular DNA loops to the nuclear matrix and are characterized by a high content of A- and T-rich regions, and clusters of topoisomerase cleavage consensus sites. MAR/SAR elements are thought to play an important role in the initiation of coordinated replication and gene expression of cellular DNA, and thus are potentially interesting targets for p53. Up to now the sequence/structure requirements for this complex but specific interaction of mutant p53 with DNA is not understood. Virtually nothing is known at the DNA and very little at the p53 protein level. Thus, at the level of the target DNA, this interaction might reflect other so far unknown p53-DNA interactions, or, alternatively, the interaction of mutant p53 with double-stranded RNA regions. The binding of MAR/SAR elements to mutant p53 is inhibited by PAb122 or PAb421, recognizing a C-terminal epitope on p53, which strongly suggests that the C-terminal nonsequence-specific DNA binding domain is involved in this interaction. However, specificity for MAR/SAR element binding seems to be conferred by amino-terminal regions of the p53 molecule, as amino-terminally truncated p53 molecules have lost the ability to specifically bind MAR/SAR elements [56]. Further experiments are aimed at understanding this activity of mutant p53, both at the structural and at the functional level. As this activity seems to be strictly mutant specific, we consider it potentially important for understanding the gain of function associated with many mutant p53 proteins.

Prospects for p53 in tumor diagnosis and therapy

The high frequency of p53 mutations in human cancer render this protein an ideal tumor marker, and as such it is already used in many clinical studies. Mutant p53 in tumor cells is present in enhanced levels, facilitating its detection in tumor tissue by immunocytochemical methods. Furthermore, monoclonal antibodies with a prevalence for p53 in a mutant conformation exist (e.g., PAb 240), already providing some additional information on the conformational status of the p53 expressed. This is becoming increasingly important, as cases accumulate, where wild-type p53 was found to be overexpressed in certain tumors [57]. However, p53 might provide more opportunities. Experiments have been initiated to probe the prognostic value of p53 analyses of various human tumors. Immunohistochemical studies seem to indicate that elevated p53 expression might be a marker for a more aggressive nature of the tumor [58]. So far, however, the prognostic power of p53 expression seems to be weak. The central issue will be, whether enhanced p53 expression indeed is an important feature of a specific malignant phenotype, playing a key role in tumor behavior. This issue is not yet resolved, but considering the different nature of the p53 mutations, and their functional consequences, this current uncertainty is not surprising. Over-expression of p53 as such might be too unspecific a criterion for a fine diagnosis of a tumor and thus for prognosis. Therefore, it will be necessary to characterize individual p53 mutations at the level of the gene and to correlate this information with diagnostic and prognostic parameters. Clearly, this is a major task, and even with the advancement of PCR-based new sequencing strategies, this approach can hardly be used for routine diagnostics. However, such information might lead to the development of more refined, e.g., peptide-based monoclonal antibodies, detecting specific subclasses of p53 mutants. Such subclass specific antibodies then would be applicable for routine diagnostics.

What about the therapeutic potential of p53? The finding that reintroduction of p53 into tumor cells can reverse the tumorigenic phenotype of p53 negative tumor cells spurred much hope for the use of wild-type p53 expression vectors in a successful somatic gene therapy. Indeed, the fact that wild-type p53 in such cells is able to overcome all adverse effects of other activated oncogenes like *ras* or *myc*, also expressed in these cells, is remarkable. Nevertheless, reasonable strategies for reintroduction and appropriate control of expression of wild-type p53 into tumors, especially into solid tumors, are far away and much more knowledge on the biology of p53 will be needed before such a strategy could be actively pursued. Nevertheless, these reintroduction studies clearly demonstrated that the tumorigenic phenotype of a cell can be permanently reverted. So the major therapeutic potential of these studies might be in the understanding of this reversal at the molecular level, perhaps allowing the design of drugs specifically mimicking the effect(s) of reintroduced wild-type p53.

Another strategic avenue for a p53-based therapeutic approach might result from the rediscovery that overexpressed, mutated p53 is a cellular tumor antigen. The fact that a certain percentage of sera from tumor patients display p53 antibodies shows that mutated (or even overexpressed) p53 is immunogenic. If it would be possible to induce a cellular immune response against the mutated p53, strategies for an immunological therapy of tumors displaying mutated p53 could be envisioned.

The ubiquitous nature and the high frequency of p53 mutations in human cancer strongly suggest that p53 is one of the key molecules in the development of human cancer. Therefore, a better understanding of p53 functions, wild-type and mutant, at the molecular level not only will further our knowledge about the biological role of this enigmatic protein, but will hold great promise for improvement of both cancer diagnostics and cancer therapy.

Acknowledgements

I thank Martina Hintz-Malchow for expert help in preparation of this manuscript, and Drs Frank Grosse and Horst-Werner Stürzbecher for helpful discussion. Research in the laboratory discussed in this review was funded by the Deutsche Krebshilfe (Dr. Mildred Scheel Stiftung). The Heinrich-Pette-Institut is financed by Freie und Hansestadt Hamburg and by Bundesministerium für Gesundheit.

References

1. Moran E. Curr Biol 1991;1:281–283.
2. Boyd JA, Barrett JC. Mol Carcinogenesis 1990;3:325–329.
3. Birchmeier W, Behrens J, Weidner KM, Frixen UH, Schipper J. Curr Opin Cell Biol 1991;3: 832–840.
4. Hollstein M, Sidransky D, Vogelstein B, Harris CC. Science 1991;253:49–53.
5. Caron de Fromentel C, Soussi T. Genes Chromosome Cancer 1992;4:1–15.
6. Lane DP, Crawford LV. Nature 1979;278:261–263.
7. Crawford L. Int Rev Exp Pathol 1983;25:1–50.
8. Mercer WE, Avignolo C, Baserga R. Mol Cell Biol 1984;4:276–281.
9. Deppert W. In: Lother H, Dernick R, Ostertag W (eds) NATO Aso series 1989, vol H34, 1989: 1701–1706.
10. Michalovitz D, Halevy O, Oren M. J Cell Biochem 1991;45:22–29.
11. Finlay CA, Hinds PW, Levine AJ. Cell 1989;57:1083–1093.
12. Eliyahu D, Michalovitz D, Eliyahu S, Pinhasi-Kimhi O, Oren M. Proc Natl Acad Sci USA 1989; 86:8763–8767.
13. Mercer WE, Shields MT, Amin M, Sauve GJ, Appella E, Romano JW, Ullrich SJ. Proc Natl Acad Sci USA 1990;87:6166–6170.
14. Michalovitz D, Halevy O, Oren M. Cell 1990;62:671–680.
15. Mowat M, Cheng A, Kimura N, Bernstein A, Benchimol S. Nature 1985;314:633–636.
16. Chow V, Ben-David Y, Bernstein A, Benchimol S, Mowat M. J Virol 1987;61:2777–2781.
17. Munroe DG, Rovinski B, Bernstein A, Benchimol S. Oncogene 1988;2:621–624.
18. Mulligan LM, Matlashewski GJ, Scrable HJ, Cavenee WK. Proc Natl Acad Sci USA 1990;87: 5863–5867.
19. Levine AJ, Momand J, Finlay CA. Nature 1991;351:453–456.
20. Soussi T, Caron de Fromentel C, May P. Oncogene 1990;5:945–952.
21. Hollstein M, Sidransky D, Vogelstein B, Harris CC. Science 1991;253:49–53.
22. Levine AJ, Momand J, Finlay CA. Nature 1991;351:453–456.
23. Gannon JV, Greaves R, Iggo R, Lane DP. EMBO J 1990;9:1595–1602.

24. Lehman TA, Bennett WP, Metcalf RA, Welsh JA, Ecker J, Modali RV, Ullrich S, Romano JW, Appella E. Cancer Res 1991;51:4090—4096.
25. Lane DP, Gannon J. Cell Biol Int Rep 1983;7:513—514.
26. Steinmeyer K, Deppert W. Oncogene 1988;3:501—507.
27. Oberosler P, Hloch P, Ramsperger U, Stahl H. EMBO J 1993;12:2389—2396.
28. Kern SE, Kinzler KW, Baker SJ, Nigro JM, Rotter V, Levine AJ, Friedman P, Prives C, Vogelstein B. Oncogene 1991;6:131—136.
29. El-Deiry WS, Kern SE, Pietenpol JA, Kinzler KW, Vogelstein B. Nature Genet 1992;1:45—49.
30. Foord O, Navot N, Rotter V. Mol Cell Biol 1993;13:1378—1384.
31. Hupp TR, Meek DW, Midgley CA, Lane DP. Cell 1992;71:875—886.
32. Friedman PN, Chen X, Bargonetti J, Prives C. Proc Natl Acad Sci USA 1993;90:3319—3323.
33. Prives C, Manfredi JJ. Genes Devel 1993;7:529—534.
34. Donehower LA, Harvey M, Slagle BL, MacArthur MJ, Montgomery CA, Butel JS, Bradley A. Nature 1992;356:215—221.
35. Maltzman W, Czyzyk L. Mol Cell Biol 1984;4:1689—1694.
36. Lane DP. Nature 1992;358:15—16.
37. Lane DP. Nature 1992;359:486—487.
38. Lane DP. Curr Opin Immunol 1992;2:581—583.
39. Prives C. Curr Opin Cell Biol 1993;5:214—218.
40. Perry ME, Levine AJ. Curr Opin Genet and Devel 1993;3:50—54.
41. Pietenpol JA, Vogelstein B. Nature 1993;365:17—18.
42. Mercer WE, Nelson D, DeLeo AB, Old LJ, Baserga R. Proc Natl Acad Sci USA 1982;79:6309—6312.
43. Deppert W, Buschhausen-Denker G, Patschinsky T, Steinmeyer K. Oncogene 1990;5:1701—1706.
44. Shohat O, Greenberg M, Reisman D, Oren M, Rotter V. Oncogene 1987;1:277—283.
45. Reich NC, Levine AJ. Nature 1984;308:199—201.
46. Milner J. Curr Opin Cell Biol 1991;3:282—286.
47. Otto A, Deppert W. Oncogene 1993;8:2591—2603.
48. Wu X, Bayle JH, Olson D, Levine AJ. Genes Devel 1993;7:1126—2232.
49. Oliner JD, Pietenpol JA, Thiagalingam S, Gyuris J, Kinzler KW, Vogelstein B. Nature 1993;362:857—860.
50. Oliner JD, Kinzler KW, Meltzer PS, George DL, Vogelstein B. Nature 1992;358:80—83.
51. Finlay CA. Mol Cell Biol 1993;13:301—306.
52. Zauberman A, Barak Y, Levy N, Oren M. EMBO J 1993;12:2799—2808.
53. Shaulsky G, Goldfinger N, Rotter V. Cancer Res 1991;51:5232—5237.
54. Dittmer D, Pati S, Zambetti G, Chu S, Teresky AK, Moore M, Finlay C, Levine AJ. Nature Genet 1993;4:42—46.
55. Milner J, Medcalf EA. Cell 1991;65:765—774.
56. Weißker S, Müller B, Homfeld A, Deppert W. Oncogene 1992;7:155—167.
57. Rodrigues NR, Rowan A, Smith MEF, Kerr IB, Bodmer WF, Gannon JV, Lane DP. Proc Natl Acad Sci USA 1990;87:7555—7559.
58. Ostrowski JL, Sawan A, Henry L, Wright C, Henry JA, Hennessy C, Lennard TJW, Angus B, Horne CHWJ. Pathol 1991;164:75—81.

Prospects in diagnosis and treatment of breast cancer
M. Schmitt et al., editors

Recent advances in studies of gene expression in clinical breast material

Gert Auer[1]*, Bo Franzén[1], Elina Eriksson[1] and Martin Bäckdahl[2]

[1]*Institute of Pathology and* [2]*Department of Surgery, Karolinska Hospital and Institute, S-171 76 Stockholm, Sweden*

Abstract. High malignant tumor variants are characterized by distinct genetic instability and elevated metastatic capacity. In this work elevated expression levels of heat shock protein 90 (hsp90) and β-tubulin (βT) in high malignant human breast cancer, analyzed by means of a combination of nonenzymatic sample preparation and two-dimensional polyacrylamide gel electrophoresis (2-DE) is reported. The data indicate a possible correlation between hsp90 and βT overexpression and clinical tumor aggressiveness in breast carcinoma and represent one step towards the identification of relevant and reliable, disease-related alterations in 2-DE patterns.

Introduction

Diagnosis of premalignant and malignant lesions are mainly based on subjective judgement of cell and tissue morphological features. Starting 20–30 years ago, objective cellular variables, e.g., crude DNA or protein contents, were increasingly measured in clinical tumor material [1]. Comprehensive studies in a variety of premalignant and malignant lesions have shown that nuclear DNA content greatly contributes to diagnostic and prognostic information above and beyond that which is obtainable by clinical and morphological parameters [2,3]. Revolutionary progress in the field of molecular biology has resulted in new knowledge concerning tumorigenesis. During the last 10 years numerous specific cellular alterations have been demonstrated to be involved in malignant transformation, and a rapidly increasing number of potential tumor markers are currently made available which can be helpful to the clinical pathologist in diagnostic decision making.

It is widely accepted that malignancies develop through a sequence of cellular events involving specific genetic alterations, genetic instability, dysregulated growth control, cell immortalization, invasive behavior and finally, metastatic potential. These distinct cellular alterations are only roughly correlated to morphological features and therefore premalignant or even malignant cells cannot be distinguished morphologically from cells exhibiting nonneoplastic changes. Thus, there is a need for objective

---

Address for correspondence: Gert Auer, Institute of pathology, Karolinska Hospital and Institute, S-171 76 Stockholm, Sweden.

markers which are complementary to modern morphological tumor diagnostic procedures. Regarding prognostic information, the situation is even worse since morphological observations are generally too basic to serve as tools for the determination of malignancy potential of a tumor in the individual patient. On the other hand, morphological methods of high quality, including tissue specific immunohistochemical stains, are frequently needed to guarantee sample specificity.

We have performed cytophotometric DNA analysis on a large number of different tumors over the past 20 years [4] and have found that the nuclear DNA content can be of value both as a diagnostic and prognostic marker. In regards to prognosis, we have found that many tumor types, e.g., breast cancer and prostate cancer, can be separated into two major categories with respect to DNA content: "diploid" or "tetraploid" tumors, and highly aneuploid tumors. Highly aneuploid tumors progress rapidly and may cause death within months to a few years, while "diploid" or "tetraploid" tumors progress slowly and could be fatal only after many years. The ploidy pattern is detected early in tumor development and generally remains unchanged during tumor progression, in other words, "diploid" and "tetraploid" tumors rarely convert to aneuploid tumors even during very long periods of observation [5].

The role of aneuploidy for malignant transformation, and the reason why aneuploid tumors are more malignant than euploid tumors remains to be clarified. The clearly deviating and variable cytophotometric DNA pattern in aneuploid tumors may reflect a high degree of genomic instability. This instability may itself lead to a generation of new phenotypes that could be a prerequisite for a rapid progression of the tumor disease. A genetic instability of this kind would reflect increased instability at the karyotypic level. The tumor cells could, for example, have acquired the ability to gain and lose chromosomes more readily. This process would favor development of the specific karyotypes required to initiate the various stages in the progression of tumorigenesis. Alternative hypotheses, including increased mutability, higher frequency of recombination events, as well as gene amplification or other DNA rearrangements, would have to be considered [6].

Two-dimensional polyacrylamide gel electrophoresis (2-DE) is a powerful, biochemical separation technique for the mapping of large numbers of known and unknown (the majority) polypeptides simultaneously [7] and is an ideal tool for the detection and identification of potential markers [8,10]. For instance, 2-DE has been used to identify differences in polypeptide expression between normal and transformed cells [8,9,11,12]. The analytical potential of 2-DE depends upon the quality of the separation and in turn upon the sample preparation technique. However, applying 2-DE to the analysis of human tumor material is not trivial:

— Frozen tumor samples may contain variable admixtures of serum proteins and connective tissue which obscure the patterns and impair the quality of the gels [13].
— Enzymatic extraction of tumor cells followed by separation using Percoll® gradient centrifugation has been used [14] to increase the quality of 2-DE analysis. However, a number of alterations have been found in 2-DE patterns as a

consequence of the enzymatic treatment [15].

We therefore developed a nonenzymatic sample preparation (NESP) technique for clinical tumor material from breast and lung [15]. One advantage with this technique is that preferentially tumor cells are extracted from the tumor tissue since they are less attached as compared with normal cells. Using NESP and 2-DE, we recently detected a polypeptide which seems to be specifically overexpressed in primary adenocarcinomas of the lung [16]. In the present report, these techniques were applied using clinical breast tumor material of potentially low and high malignant types. Our results indicate that the combination of NESP and 2-DE is a powerful way to investigate alterations in gene expression in clinical tumor samples.

Materials and Methods

Cell lines

Three breast cell lines were used as references, including one normal diploid, nontumorigenic, ER negative, fibroblast-like cell line, Hs 587Bst, and two tumorigenic, epithelial-like cell lines; MCF-7, which is "tetraploid", well differentiated and ER positive, and MDA-231, which is aneuploid, poorly differentiated and ER-negative. All cell lines were cultured as recommended by American Type Culture Collection. Cell monolayers were washed three times with PBS, harvested using a rubber policeman, and frozen as cell pellets at −70°C prior to preparation for 2-DE as previously described [15,17].

Preparation of tumor tissue samples

All specimens were obtained during 1991 and 1992 from the Department of Surgery at the Karolinska Hospital. Representative and macroscopically nonnecrotic tumor tissues were excised directly after resection, put on ice and further processed within 20 min in the presence of protease inhibitors. Cells were scraped from the surface of the tumor material using a scalpel and processed using a nonenzymatic sample preparation technique, as described previously [15]. Cell pellets from each sample were stored for 1—7 days at −70°C, and then subsequently prepared for 2-DE, as previously described [15]. As a morphological control, representative adjacent pieces from each tumor, were routinely taken for fixation in 4% buffered formalin solution, paraffin embedded, sectioned and stained using Hematoxylin eosin (HTX).

Separation of polypeptides using 2-DE

The 2-DE technique used was as described elsewhere [15,18]. Briefly, 20—30 μg of protein from each sample was used for isoelectric focusing (IEF), which was carried out in 1.2 × 180 mm rod-gels containing Resolyte 4—8 (BDH) as carrier ampholyte. The second dimension SDS-PAGE polyacrylamide concentration was 10% T.

Polypeptides were visualized by silver staining [19], and photographed with the acidic side of the gel to the left. Each sample was analyzed, at least in duplicate, by 2-DE. The level of expression of 22 polypeptides of known identity [15] were estimated using visual inspection.

Analysis of ploidy

Flow cytometry and image cytometry were used for ploidy analysis and estimation of proliferation index of clinical materials and cell lines. These methods have been described in detail previously [20].

Immunohistochemistry

Three antibodies were used for immunohistochemical characterization of formalin fixed and paraffin embedded clinical materials:
— The c-erbB-2 antibody (polyclonal) OA-11-854 (Cambridge Research Biochemicals), which was used as described elsewhere [21].
— The Ki-67 clone MIB-1 monoclonal antibody (IMMUNOTECH S.A.) was used as recommended by the manufacturer [22].
— The p53 monoclonal antibody DO-7 (DAKO), was also used as recommended by the manufacturer and included microwave treatment of the sections as described by Shi et al.[23].

Results

Clinical material

Twenty-two samples were successfully analyzed by 2-DE. However, nine samples were excluded, mainly because of polyploidy/mixed populations (as revealed by DNA ploidy analysis) or because the patient had been given chemotherapy preoperatively. It is likely that these kinds of samples produce 2-DE patterns of higher complexity, and they were therefore avoided. In this report, we selected:
— one intraductal hyperplasia without atypia, representing benign, diploid material;
— one tubular carcinoma, representing low malignant, "tetraploid" tumor material
— one comedo carcinoma, representing high malignant, aneuploid tumor material.
All three samples were c-erbB-2 negative.

Expression of polypeptides in benign vs. malignant cells

One group of actin binding proteins, the tropomyosins (TMs), have a stabilizing effect on the actin filaments. The nonmuscle tropomyosin isoforms 1—3 (TMs) were previously shown to be transformation sensitive in vitro [11,12]. We observed expression of all five TM isoforms in 2-DE gels representing intraductal hyperplasia

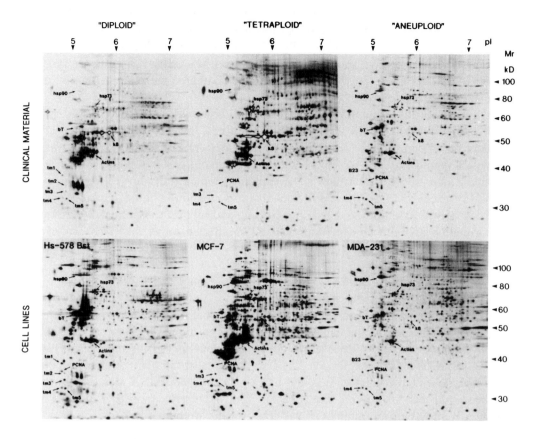

Fig. 1. 2-DE patterns of polypeptides expressed in breast cell lines and clinical material. The figure shows 2-DE analysis of: — clinical breast material (upper row): diploid intraductal hyperplasia (left), tetraploid tubular carcinoma (middle) and aneuploid comedo carcinoma (right), — breast cell lines (lower row): diploid Hs-578 Bst (left), tetraploid MCF-7 (middle) and aneuploid MDA-231 (right). The polypeptides indicated are: heat shock proteins 90 (hsp90) and 73 (hsp73), β-tubulin (βT or bT), cytokeratin 8 (k8), actins, proliferating cell nuclear antigen (PCNA), numatrin (B23) and tropomyosin isoforms 1-5 (tm1-5). The isoelectric points (pI) and molecular weights (Mr) are indicated at the top, and to the right of the figure, respectively.

and the diploid cell line Hs-578 Bst (Fig. 1). However, TM 2 was not detected in samples of tubular and comedo carcinomas, nor in MCF-7 or MDA-231 cell lines.

Expression of polypeptides in tetraploid vs. aneuploid cells

The comedo cancer and cell line MDA-231 represents the most malignant subtypes among breast malignancies. Both samples were highly p53 positive and showed a high degree of proliferation, with approximately 40% of the cells in the cell cycle, as revealed by ICM. As many as 80% of the cells in the comedo carcinoma was immunopositive for MIB-1 staining, and more than 50% of the cells were p53 positive. As expected, we observed high expression of the proliferating cell nuclear antigen (PCNA) [24] and numatrin/nucleolar protein B23 [25] in 2-DE gels from these samples (Fig. 1). In contrast, tubular carcinoma showed only intermediate p53 positivity (17% positive cells) and proliferation (20% MIB-1 positive cells). The intensity of the PCNA spot was much weaker as compared with the aneuploid tumor, and B23 was not detected.

High levels of expression of hsp90 and β-tubulin (βT) were observed in the comedo carcinoma as well as in all of the cell lines, but were weakly expressed, or undetectedable in tubular carcinoma and intraductal hyperplasia (Fig. 1). So far, we have compared six euploid tumors with six aneuploid tumors. Both βT and hsp90 showed increased expression in all aneuploid samples (not shown). On the other hand, cytokeratin 8 (k8) showed low levels of expression in comedo carcinoma, but increased expression in intraductal hyperplasia and tubular carcinoma. We observed a correlation between k8 expression and differentiation rather than between k8 and ploidy.

Expression of polypeptides in cell lines vs. clinical material

The expression levels of hsp90, βT, k8, PCNA, B23 and TMs are very similar in the comedo carcinoma and the cell line MDA-231. The tubular carcinoma and MCF-7 cells seem to express similar levels of k8 and TMs but very different levels of hsp90, βT and PCNA. The expression of TMs were also similar in the intraductal hyperplasia and Hs-578 Bst cells, but hsp90, βT, k8 and PCNA were expressed at increased levels in the cell line (Fig. 1). The expression of hsp73 did not vary markedly in any of the samples or cell lines in this study.

Discussion

Many questions remain to be answered regarding molecular events behind the highly malignant behavior of certain tumors. Experiments using cells in culture may give us some answers, however, it has been suggested that cell lines may not always be adequate models for studies of the patterns of gene expression. Adaption of tumor cells to tissue culture conditions often lead to alterations in gene expression.

We believe that a very powerful technique for the analysis of gene expression in clinical material is provided by using a combination of nonenzymatic sample preparation and 2-DE. So far, we have analyzed more than 20 different cell lines and more than 60 different tumors using 2-DE. In general, the expression of βT was increased in all cell lines compared with the tumors, and the levels of hsp90 was higher in almost all cell lines, with the exception of diploid fibroblasts. Aneuploid breast tumors (n = 6) and small cell lung carcinomas (n = 3) [16] — representing tumors with high genetic instability and metastatic capacity — showed high levels of βT and hsp90. Others have observed an increase in βT expression [8,11] during SV40 transformation of normal cells in culture. Hsp90 may be induced by a variety of treatments including heat shock [26] and transformation, using for example Ha-*ras*-oncogene [27].

Considerable knowledge has been accumulated about microtubules (MTs) and hsp90, however, little is known regarding MTs and hsp90 in high malignant tumors as compared with low malignant tumors. It is well known that the distribution of chromosomes during mitosis depends on a precise and coordinated cellular machinery, and that chromosome malsegregation can lead to chromosome loss or gain. Among the most important components of the chromosome segregational apparatus, or spindle apparatus, are MTs. These polymers are tubular organelles containing tubulins and a number of microtubule associated proteins (MAPs). Numerous drugs are known to affect the assembly of MTs (e.g., colchicine and vinca alkaloids) and these are sometimes useful as chemotherapeutic agents [28]. Several other diverse cellular functions depend upon a functioning MT-system. For instance, MTs located in the cytoplasm link various cellular organelles and membranes together and are involved in the transport of a number of substances, as well as the movement of the whole cell.

MTs were recently found to be colocalized with hsp90 [29], which also seems to be a protein having many diverse cellular functions. For instance, it has been suggested that hsp90 may serve as an antigen-presenting molecule, mediating immunity to tumors [30]. Furthermore, hsp90 may regulate DNA-binding activities of progesterone receptors in breast cancer cells [31]. Recently, a study on breast cancer showed that patients expressing high levels of hsp89 α mRNA had a significantly worse 4- and 6-year overall survival, compared with patients expressing low levels [32].

Taken together, it seems that high levels of βT and hsp90 may be a part of the high malignant phenotype. However, the precise roles of these proteins in tumor development remain to be investigated. Both polypeptides exert diverse functions, and it cannot be excluded that other functions or regulation mechanisms (in addition to those mentioned above) are of central importance. In this context, the correlation between the high malignant phenotype and high metastatic capacity may be of interest: Lakshmi et al. [33] studied the relative expression of the NM23 and MTS1 genes, which are associated with metastatic behavior, and the process of tubulin polymerization (the NM23 gene product promotes tubulin polymerization whereas the MTS1 gen product has the reverse function). They found an increase of the depolymerized form of tubulin in high metastatic melanoma cells compared with low metastatic cells.

In this report, we describe the expression of two polypeptides which may be of importance for the maintenance of genetic stability and metastatic capacity and which so far have not been studied extensively in clinical breast cancer material. These polypeptides were selected from 22 polypeptides of known identity. We did not find a strong correlation between the expression of any of the remaining polypeptides and various clinical or histological parameters. In addition, the expression of several hundred other polypeptides may be studied in the future using 2-DE, and it should be stressed that most of these are of unknown identity so far. We believe that image analysis of total polypeptide expression profiles, in combination with a detailed characterization of each tumor using clinical, histological, cytochemical and immunohistochemical methods, will produce new important information. If gene products (or its post-translational modifications) appear as potential markers, it is today technically possible to perform amino acid sequence analysis from picomol quantities of protein and later on produce antibodies for clinical applications. This study represents one step towards the identification of relevant and reliable disease-related alterations in the 2-DE patterns.

Acknowledgements

We would like to thank Inga Maurin for help with histological techniques, Ulla Aspenblad for excellently performed immunohisochemistry, Birgitta Sundelin for taking care of the cultured cells, Takashi Hirano and Ken Okuzawa for valuable help in the laboratory, Marie Edelin for careful feulgen staining and ploidy analysis, Torbjörn Hagmar for computer analysis of expression profiles and Ingeborg May for skilful photographic work. The authors also thank Martin Schalling and Stig Linder for review of the manuscript. This study was supported by grants from the Swedish Cancer Society and the Cancer Society in Stockholm.

References

1. Caspersson T, Auer G, Fallenius A, Kudynowski J. Histochem J 1983;15:337—362.
2. Fallenius AG, Franzén SA, Auer GU. Cancer 1988;62:521—530.
3. Fallenius AG, Auer GU, Carstensen JM. Cancer 1988;62:331—341.
4. Auer GU, Falkmer UG, Zetterberg AD. In: JPA Baak (ed) "Manual of Quantitative Pathology in Cancer Diagnosis and Prognosis". Heidelberg: Springer, 1991;211—232.
5. Auer G, Arrhenius E, Granberg P-O, Fox C. Eur J Cancer 1980;16,1—8.
6. Zetterberg A, Auer G. In: Ploidy level and grade of malignancy in some solid tumors. Genes and Cancer, 1984;59—67.
7. O'Farrell PH. J Biol Chem 1975;250,4007—4021.
8. Celis JE, Dejgaard K, Madsen P, Leffers H, Gesser B, Honore B, Rasmussen HH, Olsen E, Lauridsen JB, Ratz G, Mouritzen S, Basse B, Hellerup M, Celis A, Puype M, Van Damme J, Vandekerckhove J. Electrophoresis 1990;12:1072—1114.
9. Wirth PJ, Luo L, Fujimoto Y, Bisgaard HC, Olson AD. Electrophoresis 1991;11:931—954.
10. Hanash SM, Baier LJ, McCurry L, Schwartz SA. Proc Natl Acad USA 1986;83:807—811.
11 Garrels JI, Franza B Jr. J Biol Chem 1989;264:5299—5312.
12. Wirth PJ, Luo L, Fujimoto Y, Bisgaard HC. Electrophoresis 1992;13:305—320.
13. Franzén B, Iwabuchi H, Kato H, Lindholm J, Auer G. Electrophoresis 1991;12;509—515.
14. Endler AT, Young DS, Wold LE, Lieber MM, Currie RM. J Clin Chem Biochem 1986;24:981—992.
15. Franzén B, Linder S, Okuzawa K, Kato H, Auer G. Electrophoresis 1993;14:457—465.
16. Okuzawa K, Franzén B, Lindholm J, Linder S, Hirano T, Bergman T, Ebihara Y, Kato H, Auer G. Electrophoresis 1993 (in press).
17. Garrels JI. J Biol Chem 1979;254:7961—7977.
18. Anderson NL. In: Two-Dimensional Electrophoresis. Operation of the ISO-DALT System. Washington D.C.: Large Scale Biology Press, 1988.
19. Morrissey JH. Anal Biochem 1981;117:307—310.

20. Falkmer UG, Hagmar T, Auer GU. Anal Cell Path 1990;2:297–312.
21. Schimmelpenning H, Eriksson ET, Falkmer UG, Rutqvist L-E, Johansson H, Fallenius A, Auer GU. Eur J Surg Oncol 1992;18:530–537.
22. Key G, Becker MHG, Duchrow M, Schluter C, Gerdes J. Anal Cell Pathol 1992;4:181.
23. Shi S-R, Key ME, Kalra KL. J Histochem Cyochem 1991;39:741–748.
24. Cellis JE, Madsen P, Celis A, Neilsen HV, Gesser B. FEBS Lett 1987;220:1–7.
25. Feuerstein N, Chan PK, Mond JJ, J Biol Chem 1988;263:10608–10612.
26. Burdon RH. Biochem J 1986;240:313–324.
27. Lebeau J, LeChalony C, Prosperi M-T, Goubin G. Oncogene 1991;6:1125–1132.
28. Dustin P. In: Microtubles. Berlin, Germany: Springer-Verlag, 1984.
29. Redmond T, Sanchez ER, Brensnick EH, Schlesinger MJ, Toft DO, Pratt WB, Welsh MJ. Eur J Cell Biol 1989;50:66–75.
30. Srivastava PK, Maki RG. Curr Top Microbiol Immunol. 1991;167:109–123.
31. DeMarzo AM, Beck CA, Onate SA, Edwards DP. Proc Natl Acad Sci USA 1991;88:72–76.
32. Jameel A, Skilton RA, Campbell TA, Chander SK, Combes RC, Luqmani YA. Int J Cancer 1992; 50:409–415.
33. Lakshami MS, Parker C, Sherbet GV. Anticancer Res 1993;13:299–304.

Prognostic significance of histological and immunohistological tumor characteristics in node-negative breast cancer patients

T. Dimpfl*, C. Assemi and T. Genz

I. Frauenklinik der Universität München, München, Germany

Introduction

Various factors influence the development and course of a malignant breast tumor, and therefore define the prognostic scenario for the condition. They reflect the biological behaviour of the neoplasm and its host. Prognostic factors not only allow definite forecasting of the clinical development following diagnosis. They are, in certain clinical situations, decisive for the choice and likely response to treatment. Comparative case control studies have highlighted a number of indicative factors, some of which are of lesser importance, while others exercise a significant influence on the life expectancy of the affected patient.

It is generally accepted that breast cancer patients with positive lymph nodes profit from an additional systemic chemotherapy. The therapy of node-negative patients still remains an important problem. We are still not able to define exactly this patient collective and thus cannot say which women require an additional therapy and which do not. Only few studies exist that attempt to answer this question. The aim of this study is to examine nodal negative women with breast cancer, who were treated during the last 15 years, in order to study prognostic factors.

Patients and Methods

In our study we investigated 178 node-negative women with primary operable infiltrating breast cancer who were operated on between 1979 and 1983. Patients were eligible in this study if they had a minimum of seven lymph nodes removed, which histologically were all negative. The medium age was 58.9 ± 12.0 years. The youngest patient was 28, the oldest 83 years old. All patients underwent either modified radical mastectomy or tumorectomy with axillary lymph node dissection. Tumorectomy was followed by postoperative radiation of the remaining breast tissue. The mean follow-up time was 90 months.

Address for correspondence: Dr. Thomas Dimpfl, I. Frauenklinik der Universität München, Klinikum Innenstadt, Maistr. 11, 80337 München, Germany.

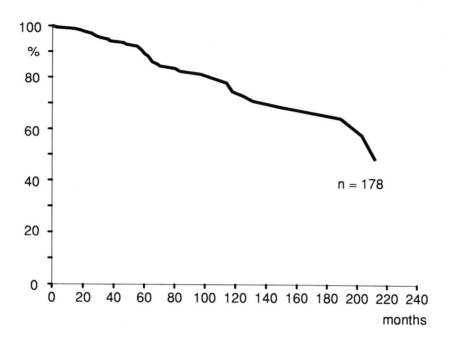

Fig. 1. Disease-free interval in 178 node-negative breast cancer patients.

As standard prognostic factors we used tumor size, histopathological classification, histological grade, blood and lymph vessel invasion and multifocality.

Of all the prognostic factors, the size of the primary tumor is the easiest to measure and costs virtually nothing. In our own lab we worked up all our specimens in so-called large-scale sections. In this way we can measure the tumor diameter very precisely.

Although the majority of mammary carcinomas (70%) are classified as infiltrating ductal carcinomas, subgroups exist with specific morphological characteristics. The classification was carried out according to the WHO recommendation, in which differentiation is primarily made between carcinoma in situ and infiltrating tumors, as well as tumors of lobular and ductal origin [1]. The histological grading was done according to the grading system of Bloom and Richardson [2].

From the so-called "new" immunohistological prognostic factors we investigated six different antigens (estrogen receptor (ER), progesterone receptor (PR), C-erB-2, p53, PCNA, Mib1) with the avidin- biotin- complex (ABC) method on formalin-fixed and paraffin-embedded material.

The interpretation of the immunohistochemical staining result was evaluated according to a semiquantitative scoring system.

All histological and immunohistochemical evaluation was done by three gynaecological pathologists (B.L., C.A., T.G.) independently of each other without knowing the clinical history of the patient.

The statistical analysis was carried out with the Statistical Package for Social

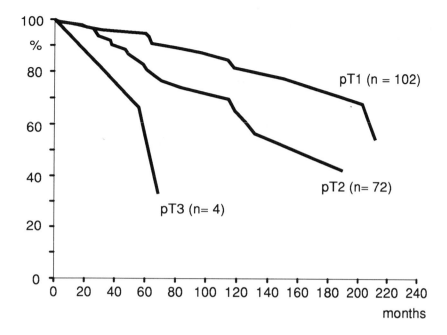

Fig. 2. Disease-free interval in node-negative patients as a function of tumor size.

Science (SPSS) software package for microcomputers: Kaplan-Meier curves were calculated for disease-free and overall survival. Stepwise multivariate analysis was

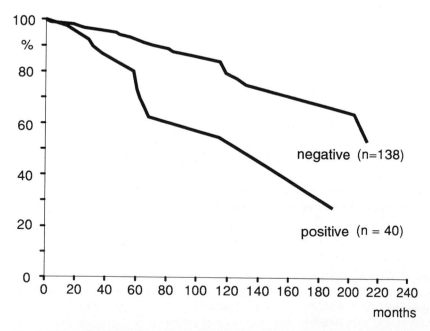

Fig. 3. Disease-free interval in node-negative patients as a function of lymphatic vessel invasion.

performed using Cox regression analysis. Disease-free survival was defined as the time between surgery and the first relapse.

Results

To characterise our 178 patients, we demonstrate the disease-free interval in Fig. 1. During surveillance 33 patients (18.5%) developed locoregional or distant recurrence. The probability to suffer recurrence 5 years after being treated for a node-negative breast carcinoma is 11%, after 10 years 25% and after 15 years 35%.

Tumor size

Women with tumors smaller than 2 cm (pT1) and negative lymph nodes had an excellent prognosis after 10 years with over 82% being disease free. The prognosis declined gradually to 65% if the tumor size was between 2 and 5 cm. It was less than 35% after 5 years for tumors larger in diameter than 5 cm (Fig. 2).

Lymphatic vessel invasion

The vessel invasion is significant for the propagation of the tumor and the further development of the disease pattern. Lymphatic invasion of cancer cells is evidence of probable dissemination and means a shorter interval of freedom from recurrence (Fig. 3).

If you analyse the above-mentioned standard prognostic factors with a Cox-regression model, it shows that in our patients only tumor size and lymphatic vessel invasion play a significant role for the development of recurrence (Table 1).

Amplification of C-erB-2 proto oncogen has been detected in 43.8% of our patients with breast carcinomas. While C-erB-2 status has not been able to predict survival in node-negative patients in most studies, a few have found a correlation. In 178 patient from our study C-erB-2 expression was analysed along with histopathology. C-erB-2 expression showed a strong correlation with early recurrence at 5 years when compared with those stained negative for C-erB-2 (Fig. 4).

Including the immunohistological parameters in the Cox regression model, you can

Table 1. Multivariate analysis for standard prognostic factors

	Significance level	Relative risk	95% confidence interval
Tumor size	0.0001	1.0506	1.03 – 1.08
Histopath. classification	0.1988		
Histopath. grading	0.2080		
Blood vessel invasion	0.9794		
Lymph vessel invasion	0.0122	2.6474	1.24 – 5.67
Multifocality	0.4546		

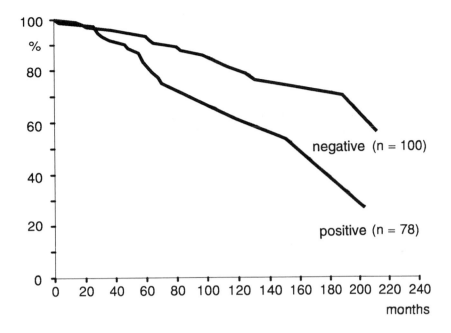

Fig. 4. Disease-free interval in node-negative patients as a fuction of C-erB-2.

see that only C-erB-2 is significant concerning the recurrence of disease (Table 2).

To gain a real statement concerning the validity of the immunohistological factors in regard to the disease-free interval, you have to analyse them together with the established standard prognostic factors. At this evaluation C-erb-2 is no longer a independent statistically significant factor, whereas tumor size and lymphatic vessel invasion remain statistically significant (Table 3).

Conclusions

Determining and using tumor characteristics that predict recurrence, survival, and response to therapy in early stage disease will become more important.

Table 2. Multivariate analysis for immunohistological factors

	Significance level	Relative risk	95% confidence interval
ER	0.1631		
PR	0.9020		
C-erB-2	0.0470	0.4860	0.24 − 1.00
p53	0.2796		
PCNA	0.2780		
Mib1	0.4761		

Table 3. Multivariate analysis for standard and immunohistological factors

	Significance level	Relative risk	95% confidence interval
Tumor size	0.0066	1.0379	1.01 − 1.06
Histopath. classification	0.2943		
Histopath. grading	0.5518		
Blood vessel invasion	0.9772		
Lymphatic vessel invasion	0.0019	3.7	1.6 − 8.4
Multifocality	0.5482		
ER	0.0886		
PR	0.9693		
C-erB-2	0.1741		
p53	0.4457		
PCNA	0.4078		
Mib1	0.8819		

At present, data indicate that tumor size and lymphatic vessel invasion can be used as indicators for the future clinical course of women with axillary node-negative disease. Other cell components of structure or function that are associated with tumor aggressiveness remain to be validated.

References

1.   WHO. Histological typing of breast tumors. 2nd edn World Health Organisation, Geneva, 1981.
2.   Bloom HJG, Richardson WW. Br J Cancer 1957;11:359–377.

New antibodies for the detection of the Ki-67 protein

Michael H.G. Kubbutat, Johannes Gerdes and Göran Key

Forschungsinstitut Borstel, Div. Molecular Immunology, Borstel, Germany

Abstract. The antibody Ki-67 has gained much attention over the last decade as a tool for the estimation of the growth fraction of human malignancies. In the last few years, there has been some progress concerning the characterization of the protein that is recognized by this antibody, although the function of this protein still remains unknown. However, a set of new antibodies could be raised which clearly recognize the Ki-67 protein even in formalin-fixed, paraffin-embedded material, thus overcoming the main drawback of the prototype Ki-67 which was only applicable in fresh frozen tissues.

Ten years ago it had been reported that the monoclonal antibody Ki-67 recognizes a protein in the nuclei of proliferating human cells [1]. Cell cycle analysis revealed that this protein is expressed in all active parts of the cycle, i.e., G_1, S, G_2 and mitosis, but not in G_0 [2]. Since then, although the nature and function of its antigen remained almost obscure, this antibody became a widely used reagent in surgical pathology and basic research [3,4]. The relevance of this antibody was constantly increasing and is still on a high level (Fig. 1).

The main drawback of antibody Ki-67 with respect to immunohistochemistry is its restriction to fresh frozen material, while it fails to react in paraffin wax-embedded tissue. During the last decade, great efforts have been made to both characterize the antigen that is defined by Ki-67 [5,6] and to generate new Ki-67-like antibodies to overcome this drawback. Such antibodies would allow a more widespread application of the estimation of the growth fraction of malignant cancers as well as extensive retrospective studies with archival material. It was not before 1991 that we could detect the Ki-67 protein in Western blots of proliferating cells as a double-band of high molecular weights (345 and 395 kD, respectively). Molecular cloning and sequencing of cDNA fragments of the Ki-67 protein revealed a first clone of 1095 bp containing repetitive sequences [7]. Thus, at least some biochemical data about this protein were obtained enabling the generation of a set of new antibodies against the Ki-67 protein. Three independent groups generated monoclonal antibodies against the Ki-67 protein by completely different strategies.

IND.64

This antibody was generated by the group of G. Delsol from Toulouse by using spleen cells from athymic nude mice grafted with Ichikawa tumour [8]. In immuno-

40

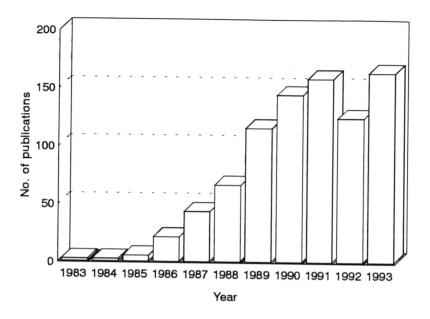

Fig. 1. Number of publications referring to antibody Ki-67. (Data from 1983–1987 taken from [3]; 1988–1993 results of plain text search on Medline on CD-ROM.)

histochemical stainings of normal tissues known to have proliferative activities, e.g., lymph node, germinal centers, small intestine and colon, IND.64 reacted with nuclei of proliferating cells. In contrast, tissues which are nonproliferative, such as kidney, cardiac muscle, liver and brain expressed no IND.64 reactivity. These findings suggested a correlation of the IND.64 antigen with cell proliferation. Staining of synchronized ABAE cells with IND.64 showed that the corresponding antigen is present in interphase cells (late G_1, S, G_2) and a strong staining was obtained in mitotic cells, whereas G_0 cells were negative. Double-labelling experiments of PHA-stimulated PBL with IND.64 and propidium iodate showed staining of about 83% of nuclei of stimulated cells after 72 h, whereas about 98% of the unstimulated cells remained negative. Thus, a similarity to antibody Ki-67 was suggested and could be substantiated by Western blot experiments with lysates of IM-9 cell line cells. IND.64 reacted with a double-band of about 345 and 395 Kd, respectively, also demonstrated by our group for Ki-67 [7]. Subsequent comparative immunobiochemical investigations with both IND.64 and Ki-67 [9] revealed a strong similarity of the two antibodies: a 1002 bp fragment and a 66 bp fragment, the latter of which was suggested to code for the Ki-67 epitope were bacterially expressed, and lysates were subjected to SDS-PAGE and Western blotting. IND.64 and Ki-67 showed identical reaction patterns with both recombinant proteins. These findings were further corroborated with a competition assay by enzyme-linked immunosorbent assay (ELISA) which showed that a synthetic peptide deduced from the 66 bp cDNA region mentioned above, was able to compete for the binding of both antibodies to a solid phase-bound antigen in a dose-dependent manner (Fig. 2).

Thus, IND.64 turned out to be a true Ki-67 equivalent with a different IgG subclass (Ki-67: IgG 1 kappa; IND.64: IgG 2b kappa) and a relative affinity which is about six times lower for IND.64 than for Ki-67 [9]. Like the prototype antibody Ki-67, IND.64 does not detect the Ki-67 protein in formalin-fixed and paraffin-embedded tissues.

The MIB "family"

By the screening of a lambda gt11 expression library with the original antibody Ki-67, a number of clones were identified which expressed a part of the Ki-67 protein [7]. Subsequent sequencing yielded the corresponding cDNA sequence which was used for the construction of expression plasmids to produce fusion proteins containing parts of the Ki-67 protein. Immunization of mice with these constructs as well as with a synthetic peptide deduced from the cDNA sequence led to a set of new monoclonal antibodies reactive with the Ki-67 protein in different test systems [10,11]. In frozen sections of human tonsils the staining pattern of antibodies MIB 1–3 was equivalent to that seen with Ki-67. This also held true for immunoblots of lysates of proliferating human cell line cells. All antibodies recognized the same high molecular weight protein double-band. The reaction in Western blots with the expression product of a 1002 bp fragment from the repetitive part of the Ki-67 protein cDNA [7,12] was identical for all antibodies. Interestingly, MIB 2 showed no staining in Western blots

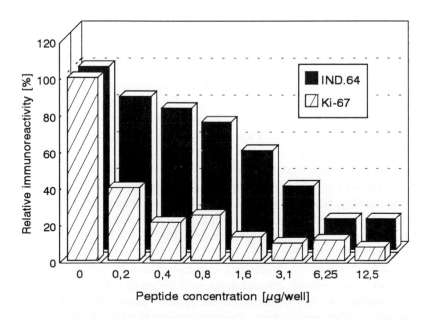

Fig. 2. Competition assay by ELISA with the antibodies Ki-67 and IND.64 and a synthetic peptide presumed to contain the epitope recognized by Ki-67 as competitor.

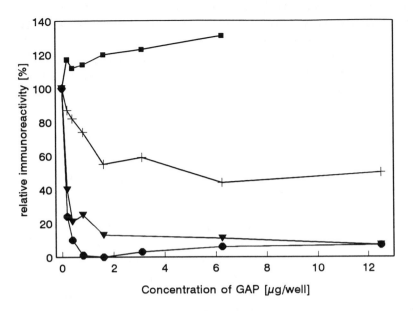

Fig. 3. Competition assay by ELISA with the new monoclonal antibodies and Ki-67, and the synthetic peptide (GAP) presumed to contain the epitope recognized by Ki-67, as competitor (▼: Ki-67; +: MIB 1; ■: MIB 2; ●:MIB 3).

of a fusion protein containing only the 66 bp element presumed to encode for the Ki-67 epitope, whereas the other antibodies also recognized this fusion protein. Competition assays with the corresponding synthetic peptide could corroborate this finding (Fig. 3). These results suggest that MIB 2 reacts with an epitope distinct from that of the other antibodies.

First attempts to detect the Ki-67 protein in paraffin sections with these antibodies were disappointing [13]. However, after microwave processing of the paraffin sections, a brilliant staining of nuclei of proliferating cells were obtained. Samples from different laboratories were tested, and the staining turned out to be independent of the formalin fixation procedure used by these groups. Even material that was fixed and embedded in 1931 was positive with MIB 1 [14].

Ki-67 and MIB 2 always failed to stain paraffin-embedded material. Interestingly, Cuevas et al. [15] reported positive staining of paraffin sections with the original antibody Ki-67. This staining, however, seemed to be dependent on strictly controlled fixation procedures, which would restrict its use in routine pathology.

Although some data are available concerning localization [5,16,17], stability [18], and molecular organization [12] of the Ki-67 protein, very little is known about the function of this protein. Besides the functional characterization of the protein by means of molecular biologic methods, the application of different antibodies can be very helpful to get insight into the behavior of the protein during the cell cycle. Recently we were able to elicit an antibody directed against the Ki-67 protein that could be of interest in this respect [12]. This antibody, MIB W21, was raised against a synthetic peptide deduced from a cDNA sequence at the 3' end of the Ki-67 protein cDNA. In Western blots of lysates of IM-9 cell line cells, MIB W21 reacts with the same double-band recognized by Ki-67 as well as with the bacterially expressed C-terminus of the protein. This provides an interesting tool for the assessment of a region of the Ki-67 protein that is apart from the central repetitive area where the epitopes of the other Ki-67 equivalent antibodies are located. For example, proteolytical modifications of this large protein during the cell cycle might be detectable by comparison of the immunohistochemical and immunobiochemical staining patterns of the different antibodies.

Rabbit antiserum against the Ki-67 protein

A rabbit antiserum was raised against a synthetic peptide deduced from the Ki-67 protein cDNA [19]. This antiserum showed reaction patterns identical to Ki-67 staining in immunocytochemistry as well as in Western blot experiments. Interestingly, this reagent is able to detect the Ki-67 protein in formalin-fixed, paraffin-embedded material after microwave irradiation of the sections, allowing the simultaneous assessment of the growth fraction and of tumour or tissue markers detected by mouse or rat monoclonal antibodies.

Ki-S5

Recently, Kreipe et al. described a new monoclonal antibody raised by immunization of mice with nuclear extracts of the Hodgkin-derived cell line L428. Preliminary experiments on Non-Hodgkin Lymphomas suggest that this antibody may be equivalent to MIB 1 [20].

Prospects

The fact that the Ki-67 protein can now be detected in paraffin material will certainly further increase its importance as operational proliferation marker. With regard to immunocytochemistry, PCNA thus far known as a cofactor of DNA polymerase δ [21,22], has been the only protein that was presumed to be related to cell proliferation and that could be detected in paraffin material [23], until McCormick and Hall recently raised some important caveats on "the complexities of proliferating cell nuclear antigen" [24]. The detection of PCNA and thus the evaluation of the results strongly depend on the fixation method and the antibody used. In a recent paper, McCormick et al. [25] described that the number of immunoreactive cells depends on the concentration of the specific antibody employed. They compared the antibodies PC10 and MIB 1 in different systems: a) an experimental tumour xenograft where the growth fraction had been previously determined by the fraction of labelled

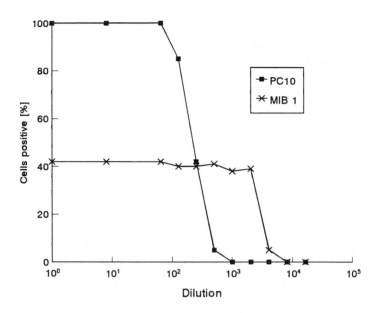

Fig. 4. Relationship of primary antibody dilution and number of positively stained cells in an experimental tumour xenograft (Data taken from [25], with alterations).

mitosis method [26] and b) exponentially growing human diploid fibroblasts. In both cases, PC10 stained 100% of cells when used neat or at low dilutions. The number of positive cells decreased with increasing antibody dilution. MIB 1, however, stained about 40% of the tumour, independent of whether the antibody was used pure or diluted. This indicates that MIB 1 staining can be applied as a reliable marker of the cell growth fraction in paraffin sections (Fig. 4).

PCNA has a relatively long half life and thus may still be present in cells that already have left the cell cycle [24]. In contrast, Ki-67 protein has a half-life of about 1 h [18] and thus readily disappears in cells in transition to G_0 [2]. Another advantage of the Ki-67 protein is that, unlike PCNA, it is not involved in DNA excision repair, as demonstrated by Hall et al. [27].

Conclusion

The immunostaining for the Ki-67 protein provides reliable means for the assessment of the growth fraction of a given human cell population. The development of new polyclonal and monoclonal antibodies, such as MIB 1, now renders this method applicable on routinely formalin-fixed, paraffin-embedded tissues, the Ki-67 protein thus being the proliferation marker of choice, e.g., retrospective studies.

Acknowledgements

This study was supported in part by the Deutsche Krebshilfe, Dr. Mildred Scheel Stiftung für Krebsforschung, project W 49/90/Ge 2.

References

1. Gerdes J, Schwab U, Lemke H, Stein H. Int J Cancer 1983;31:13–20.
2. Gerdes J, Lemke H, Baisch H, Wacher H-H, Schwab U, Stein H. J Immunol 1984;133:1710–1715.
3. Brown DC, Gatter KC. Histopathology 1990;17:489–503.
4. Gerdes J. In: Osborn M (ed) Seminars in Cancer Biology 1. London, New York: Saunders Scientific Publications, 1990;99–206.
5. Verheijen R, Kuipers HJH, Schlingemann RO, Boehmer ALM, van Driel R, Brakenhoff GJ, Ramaekers FCS. J Cell Sci 1989;92:123–130.
6. Sawhney N, Hall PA. J Pathol 1992;168:161–162.
7. Gerdes J, Li L, Schlüter C, Duchrow M, Wohlenberg C, Gerlach C, Stahmer I, Kloth S, Brandt E, Flad H-D. Am J Pathol 1991;138:867–873.
8. Meggetto F, Al Saati T, Cohen-Knafo E, Roubinet F, Selves J, Bouche G, Key G, Gerdes J, Delsol G. J Pathol 1992;168:187–196.
9. Key G, Meggetto F, Becker MHG, Al Saati T, Schlüter C, Duchrow M, Delsol G, Gerdes J. Virchows Arch B Cell Pathol 1992;62:259–262.
10. Key G, Becker MHG, Duchrow M, Schlüter C, Gerdes J. Anal Cell Pathol 1992;4:181.
11. Key G, Becker MHG, Baron B, Duchrow M, Schlüter C, Flad H-D, Gerdes J. Lab Invest 1993;68: 629–636.
12. Schlüter C, Duchrow M, Wohlenberg C, Becker MHG, Key G, Flad H-D, Gerdes J. J Cell Biol 1993; 123:513–522.
13. Gerdes J, Becker MHG, Key G, Cattoretti G. J Pathol 1992;168:85–86.
14. Cattoretti G, Becker MHG, Key G, Duchrow M, Schlüter C, Rilke F, Gerdes J. J Pathol 1992;168: 357–363.
15. Cuevas E, Jones DB, Wright DH. J Pathol 1993;169:477–478.
16. Verheijen R, Kuijpers HJH, van Driel R, Beck JLM, van Dierendonk J-H, Brakenhoff GJ, Ramaekers FCS. J Cell Sci 1989;92:531–540.
17. Isola J, Helin H, Kallioniemi O-P. Histochem J 1990;22:498–506.
18. Bruno S, Darzynkiewicz Z. Cell Prolif 1992;25:31–40.
19. Key G, Petersen JL, Becker MHG, Duchrow M, Schlüter C, Askaa J, Gerdes J. J Clin Pathol 1993; 46:1080–1084.
20. Kreipe H, Wacker H-H, Heidebrecht HJ, Haas K, Hauberg M, Tiemann M, Parwaresch R. Am J Pathol 1993;142:1689–1694.
21. Bravo R, Frank R, Blundell PA, Macdonald-Bravo H. Nature 1987;326:515–517.
22. Prehlich G, Tan CK, Kostura M, Mathews MB, So AG, Downey KM, Stillman B. Nature 1987;326: 517–520.
23. Hall PA, Levison DA, Woods AL. J Pathol 1990;162:285–294.
24. McCormick D, Hall PA. Histopathology 1992;21:591–594.
25. McCormick D, Yu C, Hobbs C. Hall PA. Histopathology 1993;22:543–547.
26. Scott RJ, Hall PA, Haldane JS, van Norden S, Price Y, Lane DP, Wright NA. J Pathol 1991;165: 173–178.
27. Hall PA, McKee PH, du P Menage H, Dover R, Lane DP. Oncogene 1992;8:203–207.

©1994 Elsevier Science B.V. All rights reserved
Prospects in diagnosis and treatment of breast cancer
M. Schmitt et al., editors

Nucleolus organizer regions (AgNORs) in ductal mammary carcinoma. Their relation to clinical factors, DNA parameters and clinical course

Michaela Aubele[1*], Gert Auer[2], Uta Jütting[1] and Peter Gais[1]

[1]GSF — Forschungszentrum für Umwelt und Gesundheit GmbH, Institut für Pathologie, Ingolstädter Landstraße 1, 85764 Oberschleißheim, Germany; and [2]Institute of Tumor Pathology, Karolinska Hospital, 10401 Stockholm 60, Sweden

Abstract. The relevance of nucleolus organizer regions (AgNORs) for classification and prognosis in breast cancer tissue was investigated. Paraffin sections from 137 cases of invasive ductal carcinomas were stained according to a modified silver staining technique and analyzed. From each case follow-up data of about 10 years (45 to 165 months), clinical, histological and several DNA distribution parameters were available. The nuclei and the silver grains were measured by means of a semiautomatic image analysis system. All resulting AgNOR parameters were investigated firstly on their correlation with the other variables and secondly on correlation with the survival time and the distant-recurrence free interval of the patients.

A significant correlation was found between most of the AgNOR parameters and the DNA features. The pTNM-stage correlated only with the standard deviation of average AgNOR area, whereas the histological grading showed highly significant correlation with several AgNOR parameters.

The prognostic significance of AgNORs was estimated using Cox regression analysis. In the multivariate approach offering all parameters available, one AgNOR feature (CV of relative AgNOR area) ranked at the third position beyond the standard deviation of DNA distribution and the pTNM-staging. Considering the distant-recurrence free interval of patients, instead of survival time, the same AgNOR feature showed an independent prognostic value.

Introduction

Nucleolar organizer regions (AgNORs) harbour genes encoding ribosomal RNA [1]. During interphase of the cells, these regions are located within the nucleolus and are associated with specifically related argyrophilic proteins [2]. They can be localized by using silver ions, which are reduced to dots of metallic silver by the argyrophilic proteins [2,3]. As these proteins are probably regulatory proteins of rDNA transcription [4] the amount of AgNORs may reflect cellular and nuclear activity. The number and size of stained dots were shown to be related to cell proliferation [5—8], cell differentiation [9], and even to malignant transformation [10—12].

---

*Address for correspondence: M. Aubele, GSF — Forschungszentrum für Umwelt und Gesundheit GmbH, Institut für Pathologie, Ingolstädter Landstraße 1, 85764 Oberschleißheim, Germany. Fax: +49-89-31873349.

After the development of a rapid one-step method of AgNOR staining by Howell and Black in 1980 [3] and its modification by Ploton [4], the utilization of silver colloid technique has been of increasing interest. The significance of AgNORs in diagnostics has, for example, been studied in melanoma [13], in lymphoma [5,14], and in breast carcinoma [11,15–18]. Studies of the predicted value of AgNORs are limited [15,19–20]. In breast carcinoma reports on the prognostic value of AgNORs are even contradictory [15,19], which may be caused by different staining methods as well as by different methods of AgNOR estimation.

The aim of our study was the evaluation of AgNORs in tissue sections of invasive ductal mammary carcinoma, by means of an image analysis system and the comparison of these values with clinico-pathological data, cytometrical DNA parameters and clinical course.

Material and Methods

Tumor samples

The material consisted of 4 μm thick paraffin sections from 137 cases, all classified as invasive ductal carcinomas without distant metastases. The material was collected at the Institute of Tumorpathology at the Karolinska Hospital, Sweden. For each case the complete data (clinical, histological, follow-up parameters) and data from cytometrical DNA measurements were available. The DNA distribution parameters had been derived in former investigations by means of image cytometry on Feulgen stained fine needle aspirates. From these, several DNA-distribution descriptors were available. In the following only those relevant for our results are described: the 2c deviation index (2c-DI) according to Böcking [21], ploidy balance [22], entropy

Table 1. Size of subgroups according to clinico-histological and follow-up data

Subgroup		No. of patients
Tumor size	pT = 1	66
	pT ≥ 2	59
Lymph node status	Negative, pN = 0	77
	Positive, pN ≥ 1	60
pTNM-stage [UICC, 26]	I	39
	II	35
	III	63
Histological grading (Bloom-Richardson)	I	12
	II	62
	III	63
Follow-up of 5 years	Alive	104
	Dead	30
	Censored	3
Distant metastases within 5 years	Yes	42
	No	95

according to Stenkvist [23], mean value (DNA_MEAN) and standard deviation (DNA_SD) of DNA distribution and ploidy (c-value of main peak). The histogram-typing was performed by an automatic DNA classification algorithm (Munich automatic Typing, MUT). There the histograms were classed according to the stemline position into diploid, tetraploid, triploid and others (multiploid or diffuse distributions).

The follow-up period was about 10 years (45–165 months). All clinical, histological and follow-up data from patients are summarized in Table 1. The patients were treated by a modified radical mastectomy [24,25]. In most of the cases subsequent irradiation was performed. Data about individual therapy differences as well as data about hormone receptor status were not available.

Staining

The AgNOR staining was performed according to the protocol of Martin [18]. Briefly, the sections were dewaxed in Xylene and then rehydrated. Prior to silver staining the Feulgen reaction was performed (Hydrolysis: 5 N HCl, 60 min. at room temperature; Schiff reagent 1 h). For silver staining the sections were then placed into a solution mixed of 2 parts of 50% silver nitrate and 1 part of 2% Gelatine in 1% aqueous formic acid. After washing off the silver colloid the slides were dehydrated and mounted.

Measurement

The investigation of the AgNORs was carried out by means of an image analysis system (Samba 2000, Alcatel, TITN Answare, Meylan, France). The measurements were performed with a 40× objective (na 0.65) and a narrow band filter of 546 nm wavelength. The resulting pixel size was $0.026 \ \mu m^2$. For each case one slide was measured by randomly selecting several fields in the peripheral and invasive areas

Table 2. AgNOR parameters calculated for each nucleus and derived features calculated per slide

Parameter	Definition
Per nucleus:	
TAAg	Total area in AgNOR
AAg	Mean area of AgNORs, TAAg/NAg
NAg	Number of AgNOR dots
ANuc	Area of nucleus
RAN	Relative area of AgNORs, TAAg/ANuc
Per specimen:	
_MEAN	Mean value
_SD	Standard deviation
_CV	Coefficient of variation
_RANGE	Maximum and minimum
_SUM	Total sum

of the tumors. At least 100 tumor cells were scanned per slide. The nuclei and the AgNOR dots were segmented automatically by a thresholding method with some degree of interactive control, in order to cut conglomerates or to reject artefacts and nuclei out of focus. In Table 2 the AgNOR features are listed, which were originally extracted and stored for each nucleus as well as distribution parameters, which were then calculated per slide, as for example the mean value (feature_MEAN), standard deviation (feature_SD) and the coefficient of variation (feature_CV).

Statistical evaluations

The statistical analyses were done using the BMDP statistical package (Statistical Software Inc, Los Angeles, CA, USA) and the SAS software (SAS Inst Inc, Cary, NC, USA). To quantify correlations of the frequency tables between different parameters the Pearson χ^2 test was used. Univariate survival probability curves were plotted using the method of Kaplan and Meier with statistical significance determined by the Wilcoxon-test. For Kaplan Meier curves stratified with continuous feature values, the data were grouped according to the 25, 50 and 25% quantitation model. Multivariate survival analysis was performed according to the stepwise Cox regression analysis. In all investigations a statistical significance was considered to be $p < 0.05$.

Results

Correlations between AgNORs and clinical, histological and cytometrical data

The axillary nodal status (pN) showed no significant correlation with any of the AgNOR parameters (Table 3A). The tumor size (pT) was correlated with the Standard Deviation (SD) of the AgNOR number (NAg_SD, p = 0.01), the pTNM-stage with the SD of the average AgNOR area (AAg_SD, p=0.02). The histological grading

Table 3A. Correlations between AgNOR parameters and the clinical data (pN,pT, pTNM-staging) and the Bloom-Richardson-grading

	pN	pT	pTNM-stage	Grading
AAg_MEAN	0.88[a]	0.52[a]	0.32[a]	<0.001
AAg_SD	0.17[a]	0.47[a]	0.02	<0.001
AAg_CV	0.93[a]	0.33[a]	0.67[a]	0.64[a]
NAg_MEAN	0.79[a]	0.23[a]	0.44[a]	0.34[a]
NAg_SD	0.39[a]	0.01	0.07[a]	0.029
NAg_CV	0.43[a]	0.24[a]	0.27[a]	0.14[a]
RAN_MEAN	0.38[a]	0.42[a]	0.54[a]	<0.001
RAN_SD	0.44[a]	0.75[a]	0.98[a]	0.1[a]
RAN_CV	0.30[a]	0.34[a]	0.31[a]	0.8[a]

p ([a] = not significant).

Table 3B. Correlations between AgNORs and the DNA distribution parameters

	Ploidy	DNA_MEAN	DNA_SD	Entropy	2c-DI	MUT
AAg_MEAN	<0.002	0.006	<0.001	<0.001	<0.001	<0.001
AAg_AD	<0.001	<0.001	<0.001	<0.001	<0.001	<0.001
AAg_CV	0.08[a]	0.19[a]	0.06[a]	0.10	0.16[a]	0.40[a]
NAg_MEAN	<0.001	<0.001	<0.001	<0.001	<0.001	<0.001
NAg_SD	<0.001	<0.001	<0.001	<0.001	<0.001	<0.001
NAg_CV	0.08[a]	0.003	<0.001	0.005	0.002	0.05
RAN_MEAN	<0.008	0.02	0.003	<0.001	0.006	0.008
RAN_SD	<0.03	<0.02	<0.002	<0.001	0.008	<0.01
RAN_CV	0.51[a]	0.90[a]	0.86[a]	0.86[a]	0.71[a]	0.40[a]

p ([a] = not significant).

correlated significantly with four AgNOR parameters, mainly those concerning the area of AgNORs (AAg_MEAN, AAg_SD, RAN_MEAN, and NAg_SD) (Table 3A). A highly significant correlation was found between the DNA distribution features and most of the AgNOR parameters (Table 3B). However, no relationship could be observed between the DNA features and the CV of average (AAg_CV) and relative AgNOR area (RAN_CV).

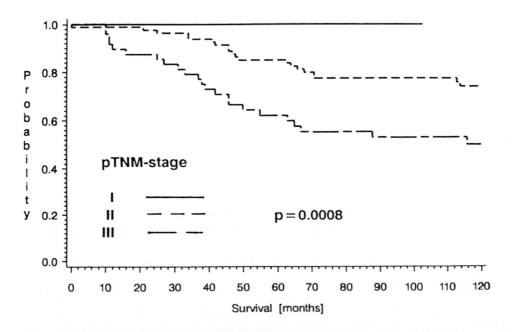

Fig. 1. Kaplan Meier survival curves for 137 ductal breast cancer patients stratified according to the pTNM-stage.

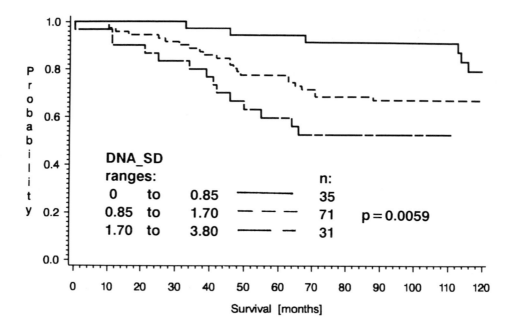

Fig. 2. Actuarial survival curves for 137 ductal breast cancer patients stratified according to the standard deviation of DNA distribution (DNA_SD) (with 25, 50 and 25% of the patients in the three groups).

AgNORs and prognosis

By means of the Cox regression analysis the prognostic value of all parameters was tested according to the survival time and the distant-recurrence free interval, respectively. A significant univariate correlation with the survival time showed the following clinico-histological and DNA parameters ($p < 0.01$): pTNM-staging, lymph node status (pN), the standard deviation of DNA distribution (DNA_SD) and the histogram type (MUT). In Figs. 1 and 2 Kaplan Meier survival curves are shown stratified according to the pTNM-stage and according to the DNA_SD. Therefore the continuous feature range of the latter was divided into three percentile groups (Fig. 2). From the AgNOR features the CV of the AgNOR number (NAg_CV), of the relative and average AgNOR area (RAN_CV, AAg_CV), and the SD of the relative AgNOR area (RAN_SD) were significant for the survival time ($p < 0.05$). Figure 3 shows as an example Kaplan Meier curves stratified according to the most significant AgNOR feature RAN_CV ($p = 0.004$). Also shown are the number of patients (n), the feature ranges per group, and the significance of differences between curves.

Testing the significance of the AgNOR parameters in a multivariate analysis all parameters including clinical, histological, DNA distribution features and AgNOR parameters were offered. The stepwise selected features were DNA_SD ($p = 0.002$), pTNM-staging ($p = 0.004$), the CV of the relative AgNOR area (RAN_CV, $p = 0.028$), and the histogram type (MUT, $p = 0.041$) (Table 4). As an illustration of this

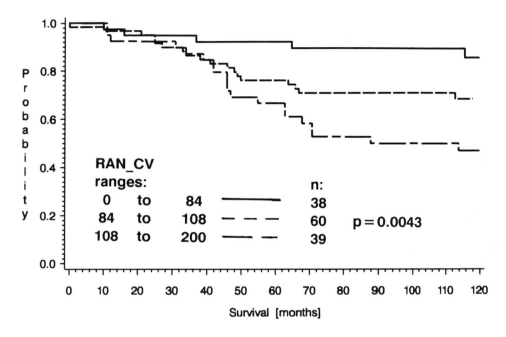

Fig. 3. Actuarial survival curves stratified according to the AgNOR feature RAN_CV (CV of relative AgNOR area) with 25, 50 and 25% quantitation of the patients.

result four patients were selected and their values were set into the formula to calculate individual survival curves. This is shown in Fig. 4. All parameters increase with worse prognosis.

In spite of the fact that pTNM-staging itself is a combined feature, we performed this analysis again and excluded the pTNM-stage. The most powerful prognostic discriminator was then DNA_SD (p = 0.002), followed by axillary nodal status pN (p = 0.01), the CV of relative AgNOR area (RAN_CV, p = 0.024) and the histogram type (MUT, p = 0.04). The fact that the impact of the second selected feature was higher for the clinical stage (p = 0.004) than for pN (p = 0.01) demonstrates a higher prognostic value of pTNM-stage.

As the occurrence of distant metastases determines the fate of the patients, we also performed Cox regression analysis for the distant-recurrence free interval. Here from the AgNOR features the CV of the relative (RAN_CV) and the average AgNOR area (AAg_CV) showed significant univariate correlation (p < 0.05). In the multivariate analysis offering all parameters again pTNM-stage and DNA_SD were proven to be the best prognostic parameters (p < 0.0001 and p = 0.032) with RAN_CV (p = 0.021) at the third position of this model. These results are also shown in Table 4.

The fact that RAN_CV is neither correlated with pTNM-stage, nor with any of the DNA features, makes it an additional prognostic factor in breast cancers.

Table 4. Results of stepwise Cox regression analyses

Selected variable	Improvement	
	χ^2	p
Total survival time (all parameters offered):		
DNA_SD	9.8	0.002
pTNM-STAGE	8.4	0.004
RAN_CV	4.8	0.03
MUT	4.2	0.04
Total survival time (pTNM-staging excluded):		
DNA_SD	10.4	0.002
pN	7.9	0.010
RAN_CV	5.5	0.024
MUT	3.8	0.04
Distant-recurrence free interval (all parameters offered):		
pTNM-STAGE	12.3	≤0.0001
DNA_SD	4.6	0.032
RAN_CV	5.3	0.021

Discussion

The development of an AgNOR staining technique for formalin-fixed, paraffin-embedded tissue opened a wide field of application. In the last few years many authors reported on the relevance of AgNORs in diagnostics on various tissues. Higher AgNOR counts were found in malignant breast lesions compared with benign ones [8,11,27]. The AgNOR count was also reported to differ significantly between breast cancer patients with axillary lymph node metastases (pN+) and those without (pN0) [15], and therefore could be considered as an indicator for good or bad prognosis of patients. However, there is also indication that exclusive counting of AgNORs does not correlate with prognosis of breast cancer patients [19]. Such contradictions may be caused by different staining methods as well as by the estimation of AgNORs (as for example when counting manually or measuring quantitatively), resulting in various parameters. As there is no consensus at present as to which AgNOR parameter might be of diagnostic or prognostic interest, we offered all our parameters to multivariate analyses. Our results show that features concerning the AgNOR area have a higher prognostic value than the number of AgNORs. Therefore image analysis in the evaluation of AgNORs is suggested to have decisive advantages over subjective evaluations. Not only because of the quantitative determination of the AgNOR area, but also because of the possibility to use more sensitive features like distribution descriptors, whose prognostic value could clearly be shown.

Traditional methods of prognostication in breast cancer include mainly clinical staging and histological grading [25,26,28,29]. Also hormone receptor status and nuclear DNA content [25,28] have been introduced in order to obtain prognostic

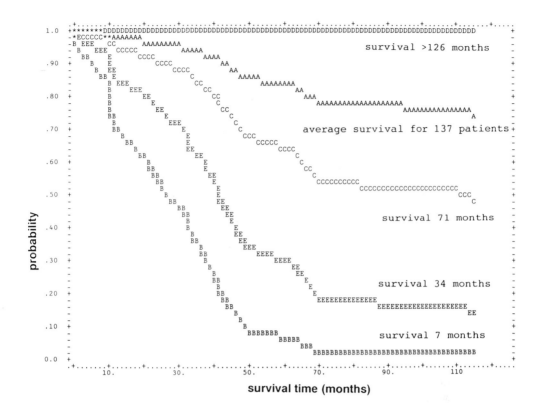

Fig. 4. Examples for individual survival probability curves for four breast cancer patients. The curves are stratified according to the most potent parameters from Cox analysis: SD of DNA distribution (DNA_SD), pTNM-stage, CV of relative AgNOR area (RAN_CV) and histogram type (MUT). The average survival probability for all 137 patients is also shown.

information and to guide treatment. The main attempt in our study was to investigate whether quantitatively measured AgNOR parameters may improve prognostication. In our results of Cox regression analysis from all parameters, the SD of DNA distribution showed the highest correlation with survival time, followed by pTNM-staging. At the third position ranked an AgNOR parameter (RAN_CV), which was proven not to be correlated with the others. Therefore, we can conclude that the described AgNOR parameters have additional prognostic relevance, and may improve and refine the quality of individual patient prognosis.

Acknowledgements

This work was supported by a grant from the "Wilhelm-Vaillant-Foundation" (München, FRG), the "Swedish Cancer Society" and the "Cancer Society of Stockholm".

The authors would like to thank Prof W. Gössner (GSF, 85764 Oberschleißheim, FRG) for advice and fruitful discussions.

We also acknowledge Prof H. Martin and his group (Humboldt Universität Berlin, FRG) and Prof K. Kunze and his group (Medizinischen Akademie Dresden, FRG) for helpful interchange of AgNOR staining experiences.

References

1. Henderson AS, Warburton D, Atwood KC. Proc Natl Acad Sci USA 1972;69:3394−3398.
2. Ploton D, Visseaux-Coletto B, Canellas JC et al. Anal Quant Cytol Histol 1992;14(1):14−23.
3. Howell WM, Black DA. Experientia 1980;36:1014−1015.
4. Ploton D, Menager M, Jeannesson P et al. Histochem J 1986;18:5−14.
5. Crocker J, McCartney JC, Smith PJ. J Pathol 1988;154:151−156.
6. Dervan PA, Gilmartin LG, Loftus BM et al. Am J Clin Pathol 1989;92:401−407.
7. Hall PA, Crocker J, Watts A et al. Histopathology 1988;12:373−381.
8. Raymond WA, Leong AS. Human Pathol 1989;20(8):741−746.
9. Rüschoff J, Plate K, Bittinger A et al. Path Res Pract 1989;185:878−885.
10. Derenzini M, Nardi F, Farabegoli F et al. Acta Cytol 1989;33:491−498.
11. Smith R, Crocker J. Histopathology 1988;12:113−125.
12. Underwood JCE, Giri DD. J Pathol 1988;155:95−96.
13. Fallowfield ME, Dodson AR, Cook MG. Histopathology 1988;13:95−99.
14. Crocker J, Paramjit N. J Pathol 1987;151:111−118.
15. Bockmühl U, Theissig F, Dimmer V et al. Path Res Pract 1991;187:437−443.
16. Charpin C, Bonnier P, Piana L et al. Path Res Pract 1992;188:1009−1017.
17. Rüschoff J, Neumann K, Contractor H et al. J Cancer Res Clin Oncol 1990;116:480−485.
18. Martin H, Hufnagl P, Beil M et al. Anal Quant Cytol Histol 1992;14(4):312−319.
19. Toikkanen S, Joensuu H. J Pathol 1993;169:251−254.
20. Böhm J, Kacic V, Gais P et al. Histochemistry 1993;99:85−90.
21. Böcking A, Chatelain R, Biesterfeld S et al. Anal Quant Cytol Histol 1989;11,2:73−80.
22. Opfermann M, Brugal G, Vassilakos P. Cytometry 1987;8:217−224.
23. Stenkvist B, Westman-Naeser S, Veggelius J et al. J Clin Pathol 1979;32:979−983.
24. Auer GU, Caspersson T, Wallgren A. Anal Quant Cytol Histol 1980;3:161−165.
25. Fallenius AG, Auer GU, Carstensen JM. Cancer 1988;62:521−530.
26. UICC (Union Internationale Contre le Cancer):TNM. Klassifikation maligner Tumoren. 4. Auflage. Berlin: Springer Verlag, 1987.
27. Giri DD, Nottingham JF, Lawry J et al. J Pathol 1989;157:307−313.
28. Bloom HJG, Field JR. Cancer 1971;28:1580−1589.
29. Champion HR, Wallace IWJ. Cancer 1971;25:441−448.

Prospects in diagnosis and treatment of breast cancer
M. Schmitt et al., editors

Patterns of keratin expression and differentiation of breast cancer

W.B.J. Nathrath[1*] and E.B. Lane[2]
[1]*Institut für Allgemeine Pathologie und Pathologische Anatomie der Technischen Universität München, Klinikum Rechts der Isar, 81675 München, Germany; and* [2]*CRC Cell Structure Research Group, Cancer Research Campaign Laboratories, Department of Anatomy & Physiology, University of Dundee, Dundee DD1 4HN, UK*

Abstract. The distribution of keratins (K) in 100 breast carcinomas was investigated, using a wide range of monoclonal mono- or oligospecific antibodies against K1/2,4,7,8,10/11,13,14,18,19. Two main staining patterns were distinguished: K8,18,19 and K7,8,18,19 reactive carcinomas. In addition, four carcinomas also expressed K14, and one carcinoma was K1/2 reactive. K7 expression was negatively correlated with the expression of estrogen receptor (ER) and with age: most carcinomas of patients under 55 years of age were K7 positive and ER negative, whereas a large group of K7 negative, ER positive carcinomas were found within the older age group.

Introduction

Using the established histomorphological classification [1], about 70% of all breast carcinomas belong to the invasive ductal group of no special type, in which a wide range of prognosis and clinico-biological behaviour can be observed. Yet subclassification of this group is difficult on the basis of histology only [2]. Apart from additional information about tumor stage and grade, recent studies have shown, or indicated the potential of certain biochemical markers to further subdivide breast carcinomas into prognostically and therapeutically different groups [3–6].

Keratins (K) may be candidates for this approach, since their profiles of expression correlate specifically with different types of epithelial differentiation and function. The 30 or more K proteins constitute the epithelial specific intermediate filaments and belong to the two largest groups, i.e., type I and type II, of the intermediate filament-like gene family proteins. At least one member of each family is expressed by any epithelium [7,8], and six coexpressed pairs of K define basic epithelial phenotypes: one primary pair (K5 + K14) and four secondary differentiation specific pairs in the complex stratified epithelia, and one primary pair (K8 + K18) for simple epithelial cells.

Normal human mammary gland has been shown to express K8 and 18, and in addition also K5,7,14,17 and 19 [7,9], of which K7,8,18 and 19 are correlated with luminal cells, whereas K5 and 14 are found in the myoepithelial (basal) cells of the

---

Address for correspondence: Prof Dr W. Nathrath, Institut für Allgemeine Pathologie und Pathologische Anatomie der TUM, Ismaninger Str. 22, 81675 München, Germany.

Table 1. Antibodies employed

Keratin specificity	Clone	Source	References
7	RPN 1162	Amersham	[16]
7,8	LP1K	Lane	[17]
7, [8]	LP5K	Lane	[17]
8	LE41	Lane	[18]
8	TS1	Stigbrand	[19]
18	LE61	Lane	[18]
18	LE65	Lane	[18]
19	LP2K	Lane	[17]
19	K 4.62	Bio-Makor	[20]
1/2	RPN 1161	Amersham	[21]
10	RKSE 60	Eurodiagnostics	[22]
10/11	K 8.60	Bio-Makor	[23]
13	KS 13.1	Progen	[24]
14	LL001	Lane	[14]
1,5,7,8,15	TS4	Stigbrand	[19]
1—8	LP3K	Lane	[17]
5,18	LP34	Lane	[15]

ductal tree [10—14]. In breast carcinomas at least two K expression patterns have been observed biochemically, i.e., type I (K7,8,18,19) and type II (K6,7,8,11,14,16,17,18, 19) [7,9]; irregular, weak or no expression of K7 in some breast carcinomas has been mentioned [9,10].

In this study, we describe the immunohistological distribution of K in a wide range of breast carcinomas and the relationship of differences in K7 expression with estrogen receptor content of carcinomas and with age of carcinoma patients.

Material and Methods

Samples from surgical biopsy specimen of 100 randomly chosen breast carcinomas, with normal gland tissue attached, were frozen in liquid nitrogen and stored at −80°C. 5μm cryostat sections were briefly fixed in ice-cold acetone. The ABC immuno-peroxidase method was applied as described previously [15]. The antibodies used are listed in Table 1.

Biochemical and immunohistological results of estrogen- and progesterone-receptor determination were available for each carcinoma.

The age distribution of the carcinoma patients was bimodal, i.e., there was a separation at 55 years, which led to two frequency distributions, with 58 patients in the older and with 42 patients in the younger age group.

The carcinoma types, diagnosed according to the WHO classification [1], showed the usual frequency distribution (Table 2).

Data of 40 patients were available for a follow-up analysis of disease-free survival over 5 years.

Table 2. Immunohistological distribution of keratins

Number	Type of carcinoma	Keratins			
		7	8 18 19	1/2 4 10/11 13	14
17	lobular invasive	12+ 5–	17+	17–	17–
69	ductal invasive (NOS)	51+ 18–	69+ 1–[a]	1+[b] 69–	3+ 66–
1	ductal in situ	1+	1+	1–	1–
7	mucinous	2+ 5–	7+	7–	7–
3	medullary	3+	3+	3–	1+ 2–
2	tubular	2+	2+	2–	2–
1	papillary	1–	1+	1–	1–
100		71+ 29–	100+	100–	4+ 96–

[a]One ductal invasive carcinoma was negative for keratin 19; [b]one ductal invasive carcinoma was positive for keratin 1/2.

Results

Normal mammary gland

Specific immunostaining for K7,8,18 and 19 was found in luminal epithelial cells, with occasional negativity for each of the four K, especially for K19. Basal/myoepithelial cells showed strong and uniform reactivity for K14; there was weak staining of some basal cells for K7 and 19. The luminal simple epithelial cells were basically negative for K14 with the exception of some single cells in preterminal and also in large ducts.

Breast carcinomas

The details of the staining distribution of K in 100 carcinomas are summarised in Table 2. All carcinomas showed homogeneous staining for K8,18 and 19 with at least one of the antibodies used. The K7 antibodies gave the following staining results: there was a large group of completely positive and a small group of completely negative carcinomas; in addition, a third group of carcinomas showed heterogeneous K7 staining. To discriminate low and high level of K7 expression in this heterogeneously staining group, a cut-off level was set at 30%, i.e. a K7 low level carcinoma was nonreactive in at least 70% of the carcinoma cells, and a K7 high level carcinoma expressed K7 in more than 30% of the carcinoma cells. When coalising these low and high level K7 carcinomas with the completely K7 negative and positive cases, respectively, 29 carcinomas were operationally considered to be K7 "negative" and 71 carcinomas to be K7 "positive".

Lobular and ductal invasive carcinomas, including mucinous, showed identical staining behaviour for the four "luminal type K" 7,8,18 and 19. While virtually all carcinomas were nonreactive for the K 4,10/11 and 13, there were one medullary carcinoma and three invasive ductal carcinomas of no special type (not otherwise specified, NOS) with strong keratin 14 expression. One of these NOS-carcinomas was reactive for K1/2. This and the other three K14 positive carcinomas also expressed K7,8,18 and 19.

Correlation of K7 staining with age of patients and with expression of estrogen receptor in carcinomas

Thirty-three of the 42 patients younger than 55 years had K7 positive carcinomas, 27 of which were ER negative; only nine carcinomas in this younger age group were K7 negative, seven of which were ER positive.

In the older age group, 20 of the 58 carcinomas were K7 negative; they all were ER positive. The 38 K7 positive carcinomas dichotomised into a group of 22 cases, which expressed both K7 and ER, and a group of 16 K7 positive/ER negative carcinomas, corresponding to the predominant staining pattern of the younger age group.

The analysis of disease-free survival of 40 patients over 5 years, using Cox's proportional hazard model, revealed an increased risk of relapse in patients with a K7 positive carcinoma.

Discussion

The results presented here are in agreement with previous results of other studies [7,9–11,13,24,25]: first, breast carcinomas can maintain the K7,8,18,19 ductal simple epithelial staining pattern of the normal mammary gland; secondly, few carcinomas show an additional K14 expression; thirdly, there is heterogeneity in K7 expression in a number of carcinomas. In our study, an additional group of carcinomas was found, which was completely negative for K7. By drawing a cut-off line at 30% for the heterogeneous K7 group and combining the accordingly low level K7 with the completely negative K7 cases, altogether 29 carcinomas were operationally considered to be "negative" for K7. K7 positive and negative cases have been found in other carcinoma groups, e.g., renal cell, hepatocellular and stomach carcinomas, and in these the K7 expression is associated with a low grade differentiation state of the carcinoma [9,26,27]. Similarly, in a number of the breast carcinomas presented here, the keratin 7 expression is associated with ER-negativity, considered to indicate low grade differentiation [3]. First follow-up data of 40 patients in our study show a significantly increased risk of carcinoma relapse within 5 years, in patients with K7 positive carcinomas. Our results also indicate that there is a preponderance of this K7 positive/ER negative carcinoma type in the young age group, probably suggesting a premenopausal carcinoma type, different from a hypothetical "old age" carcinoma type. Furthermore, four of the K7 positive carcinomas show K14 expression. K14 is the main intermediate filament, coexpressed with K5, of the myoepithelial cells [14], which also express smooth muscle actin [11,28]. K14 reactivity, without smooth muscle actin, is also found in some luminal cells of small preterminal ducts; there is some likelihood that these luminal K14 reactive cells are committed post stem cells with a transformational state towards the luminal cells [14,25,29]. These cells are comparable to developmentally early mammary duct cells, which share K5, K14 and also K19 with basal cells of epidermis, from which state they develop into the two-layered mixed duct epithelium with the keratin distinction between the luminal and basal cells [25,29]. The subpopulation of luminal cells in the small ducts of the mature mammary gland, which express the dual luminal and basal keratin phenotype, may be the origin or equivalent of a subgroup of breast carcinomas with a tendency to reactivate the early keratinocyte-like developmental state with some relationship to epidermis. This hypothetical group may be defined as to contain carcinomas which are capable of expressing K14 and other squamous epithelial differentiation K; indeed, one of our K14 positive cases showed additional strong staining for the squamous epithelial K1/2. Furthermore, the type II K profile of breast carcinomas, identified by Moll et al. [9], includes the simple epithelial K, K14 and the squamous epithelial keratin 11. The observation that about 4% of breast carcinomas show squamous epithelial alteration [30] may lend further support to the hypothesis of a "keratinocyte" type of breast carcinomas.

If further studies confirm the data given here, a keratinocyte-like differentiation is found in a subgroup of mammary carcinomas, which also express the simple luminal epithelial keratin profile 7,8,18,19, and which tend to be ER negative and to be predominant in a premenopausal patient group. A carcinoma type with the "pure" simple K profile 8,18,19 would then probably present a postmenopausal type carcinoma.

Work is in progress on a larger number of carcinomas to substantiate the data given here, by including monoclonal antibodies against remaining K types so that the cellular profile of breast carcinoma can be defined fully, and may in connection with other biochemical, morphological and clinical data allow further subdivision of breast carcinomas with impact on prognosis and therapy.

Acknowledgements

The authors would like to thank Brigitte Popp for excellent immunohistochemical work. This work was funded by the Wilhelm-Sander-Stiftung, grant no. 82007.3 (WBJN).

References

1. Azzopardi JG, Chepizk OF, Hartman WH. International Histological Classification of Tumours No.2: Histological typing of breast tumours. 2nd edn. Geneva: World Health Organisation, 1981.
2. Page DL, Anderson TJ. Diagnostic Histopathology of the Breast. Edinburgh, London, Melbourne, New York: Churchill Livingstone, 1987.
3. Cattoretti G, Andreola S, Clemente C, D'Amato L, Rilke F. Br J Cancer 1988;57:353–357.

62

4. Dawkins HJS, Robbins PD, Smith KL, Sarna M, Harvey JM, Sterret GF, Papadimitriou JM. Path Res Pract 1993;189:1233–1252.
5. Jänicke F, Schmitt M, Pache L, Ulm K, Harbeck N, Höfler H, Graeff H. Breast Cancer Res Treat 1987;10:5–9.
6. Klijn JGM, Berns PMJJ, Schmitz PIM, Foekens JA. Endocrinol Rev 1992;13:3–17.
7. Moll R, Franke WW, Schiller DL, Geiger B, Krepler R. Cell 1982;31:11–24.
8. Sun TT, Eichner R, Schermer R, Cooper D, Nelson WG, Weiss RA. In: Levine AJ, Vande Woude GF, Topp WC, Watson JD (eds) Cancer Cells/The Transformed Phenotype. Cold Spring Harbor, New York: Cold Spring Harbor Laboratory, 1984;169–176.
9. Moll R, Krepler R, Franke WW. Differentiation 1983;23:256–269.
10. Altmannsberger M, Dirk T, Droese M, Weber K, Osborn M. Virchows Arch B 1986;51:265–275.
11. Böcker W, Bier B, Freytag G, Brömmelkamp B, Jawisch ED, Edel G, Dockhorn-Dworniczak B, Schmid KW. Virchows Arch A 1992;421:323–330.
12. Franke WW, Schmid E, Freudenstein C, Appelhaus B, Osborn M, Weber K, Keenan TW. J Cell Biol 1980;84:633–654.
13. Nathrath WBJ, Wilson PD, Trejdosiewicz LK. Path Res Pract 1982;175:279–288.
14. Purkis PE, Steel JB, Mackenzie IC, Nathrath WBJ, Leigh IM, Lane EB. J Cell Sci 1990;97:39–50.
15. Nathrath WBJ, Lane EB, Peat NP, Purkis PE, Leigh IM. The pan epithelial anti-keratin monoclonal antibody LP34 recognizes epitopes on both type I and type II keratins, which are differentially lost and recovered in immunohistology by different tissue processing. J Histochem Cytochem (in press).
16. Tölle HG, Weber K, Osborn M. Eur J Cell Biol 1985;38:234–244.
17. Lane EB, Bartek J, Purkis PE, Leigh IM. In: Wang E, Fischman D, Liem RHK, Sun TT (eds) Intermediate Filaments. New York: NY Acad Sci Ann 455, 1985;241–258.
18. Lane EB. J Cell Biol 1982;92:665–673.
19. Sundström BE, Nathrath WBJ, Stigbrand TI. J Histochem Cytochem 1989;37:1845–1854.
20. Gigi-Leitner O, Geiger B. Cell Motil Cytosk 1986;6:628–639.
21. Osborn M, Altmannsberger M, Debus E, Weber K. Ann NY Acad Sci 1989;455:649–668.
22. Ramaekers FCS, Moesker O, Huysmans A, Schaart G, Westerhof G, Waagemar SS, Herman CJ, Vooijs GP. In: Wang E, Fischman D, Liem RHK, Sun TT (eds) Intermediate Filaments. New York: NY Acad Sci Ann 455, 1985;614–634.
23. Huszar M, Gigi-Leitner O, Moll R, Franke WW, Geiger B. Differentiation 1986;31:141–153.
24. Moll R. Verh Dtsch Ges Path 1991;75:446–459.
25. Taylor-Papadimitriou J, Lane EB. In: Neville MC, Daniel CW (eds) The Mammary Gland. London: Plenum Publishing Corporation, 1987;181–215.
26. Pitz S, Moll R, Störkel S, Thoenes W. Lab Invest 1987;56:642–653.
27. Van Eyken P, Sciot R, Desmet VJ. Histopathology 1989;15:125–135.
28. Draeger A, Nathrath WBJ, Lane EB, Sundström BE, Stigbrand TI. APMIS 1991;99:405–415.
29. Dulbecco R, Allen WR, Bologna M, Bowman M. Cancer Res 1986;46:2449–2456.
30. Fisher ER, Gregorio RM, Fisher B. Cancer 1975;36:1–85.

Flow cytometric DNA analysis in primary breast cancer: technical pitfalls and clinical applications

N. Harbeck[1*], N. Yamamoto[1], N. Moniwa[1], E. Schüren[1], P. Ziffer[1], P. Dettmar[2], H. Höfler[2], M. Schmitt[1] and H. Graeff[1]

[1]Frauenklinik and [2]Institut für Allgemeine Pathologie und Pathologische Anatomie, Technische Universität, München, Germany

Abstract. The prognostic impact of proliferation markers is still controversial, partly due to lack of standardization of tissue extraction techniques and data analysis. Compared to older proliferation assays (e.g., thymidine labeling index, image cytophotometry), flow cytometric DNA analysis has become increasingly popular. It provides information about both the proliferative activity (S-phase) and the DNA content (ploidy) of a tumor; both have been associated with prognosis in primary breast cancer. This article focuses on two major applications of flow cytometric DNA analysis: formalin-fixed, paraffin-embedded tissue and fresh tumor tissue. The importance of valid laboratory bench techniques as well as computer-based data analysis for clinical application in primary breast cancer is discussed.

Introduction

Metastatic capacity and proliferative activity are the two most important characteristics for describing the malignant behavior of a tumor. Thus, they offer a tumor-biological basis for new prognostic factors. Numerous techniques have been proposed for assessing a tumor's proliferative capacity. One of the most widely used methods is flow cytometric DNA analysis.

Technical applications of flow cytometry have been considerably refined within the past 30 years, from pure cell counting to sophisticated analyses of cell properties using monoclonal antibodies as well as specific fluorochromes. As early as 1969 flow cytometrically created DNA histograms were shown by Van Dilla et al. [1]. However, flow cytometric DNA analysis has only become increasingly popular over the last decade. Early reports about the proliferative potential of a tumor had been based on very labor-intensive techniques, like thymidine labeling index (TLI) [2] or image cytophotometry [3]. These papers suggested an important prognostic impact of proliferation capacity in cancer patients, but due to the very sophisticated techniques used, oncologists sought for assays that could be more conveniently performed on a routine basis. Flow cytometric DNA analysis proved to be a very feasible alternative,

---

*Address for correspondence: Nadia Harbeck MD, Frauenklinik, Klinikum rechts der Isar, Technische Universität München, Ismaningerstr. 22, D-81675 München, Germany. Tel.: +49-89-4140-2417. Fax: +49-89-4180-5146.

facilitating the processing of large cell numbers in a very short time. In addition, it is able to provide information about the proliferative potential (S-phase) as well as the DNA content (ploidy) of tumor cells. Flow cytometric DNA analysis in fixed, paraffin-embedded tissue [4] enabled retrospective studies of archived tumor specimens. Thus, it became possible to evaluate S-phase and ploidy as potential prognostic factors for survival in large patient collectives. Although numerous additional methodological advances have recently been made, the prognostic impact of flow cytometrically determined S-phase and ploidy is still controversial, as the lack of standardization in tissue extraction and differing data evaluation make comparison of results by different research groups very difficult. This precludes routine application of DNA flow cytometry in small institutions.

Nevertheless, reliable and reproducible laboratory assays are essential for testing the prognostic relevance of any tumor biological factor for clinical application. This article will focus on the technical aspects of flow cytometric DNA analysis, emphasizing applications in primary breast cancer; it will outline methodological pitfalls and discuss new technical modifications.

Flow cytometric DNA analysis in paraffin-embedded tissue

In 1983, Hedley et al. [4] presented a new technique for preparation of nuclei from formalin-fixed, paraffin-embedded tissue sections involving pepsin as the releasing agent, which has been widely used ever since in flow cytometry. This approach enabled the preparation of nuclei from paraffin-embedded tissue for the flow cytometric analysis of DNA. DNA analysis in such specimens had previously only been possible for image cytometry on Feulgen-stained sections.

Nuclei released from paraffin-embedded sections by proteolytic tissue degradation should be cleared of cytoplasmic and surface antigens, thus rendering nuclear size and granularity the predominant markers for identification of different populations in flow cytometry. This constitutes one of the major setbacks of Hedley's original method: In addition to "bare" nuclei, it also results in a quite heterogeneous suspension consisting of cell debris and nuclei surrounded by cytoplasmic residues. This heterogeneity makes proper identification of nuclei by flow cytometry more difficult and computer-based analysis of the DNA histogram is often impossible. These difficulties are even more pronounced in a tumor consisting of tumor cells as well as stromal cells such as lymphocytes, fibroblasts, and phagocytic cells.

Our group has modified Hedley's method, resulting in a preparation of bare, homogeneous nuclei from formalin-fixed, paraffin-embedded breast cancer sections [5]. This homogeneity permits proper identification of nuclei, cuts down debris and therefore facilitates computer-based data analysis (Fig. 1). For each patient, only one 50 μm formalin-fixed, paraffin-embedded tissue section containing at least 80% tumor cells (pathologist's report) is required. After deparaffinization and rehydration, proteolytic digestion is performed. The major modification step is an increase of pepsin digestion time which completely destroys the surrounding cytoplasmic

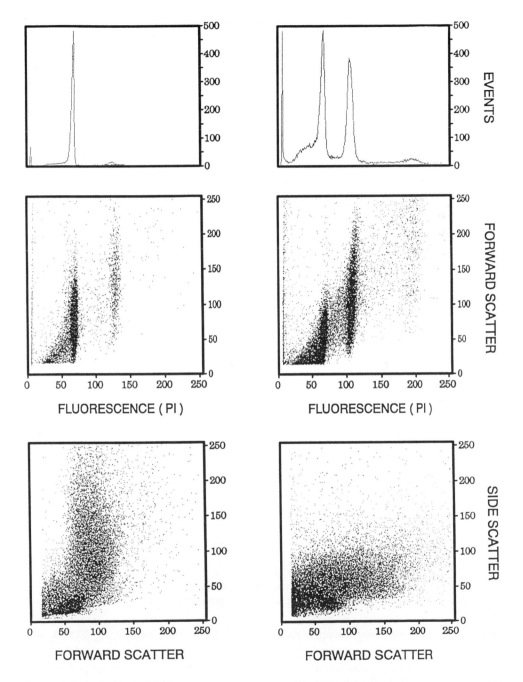

Fig. 1. Flow cytometric analysis of formalin-fixed, paraffin-embedded breast cancer samples: diploid tumor (left) and aneuploid tumor (right).

66

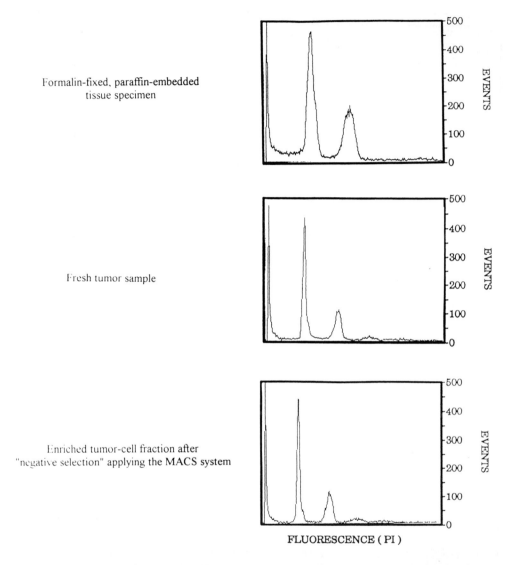

Formalin-fixed, paraffin-embedded tissue specimen

Fresh tumor sample

Enriched tumor-cell fraction after "negative selection" applying the MACS system

FLUORESCENCE (PI)

Fig. 2. Flow cytometrically determined DNA content in different tissue specimens of the same ovarian carcinoma: Formalin-fixed, paraffin-embedded tissue specimen (upper plot), fresh tumor sample (middle), and enriched tumor-cell fraction after "negative selection" applying the MACS system (lower).

constituents. The resulting nuclei suspension is put into PBS containing EDTA and RNAse, which minimizes aggregation during storage. Nuclei are then stained for flow cytofluorometry with propidium iodide, a fluorescent dye that specifically binds to double-stranded DNA. Results obtained with this modified Hedley's technique proved to be highly reproducible, and preparations from paraffin-embedded tissue sections appeared quite similar to those obtained from fresh tumor samples (Fig. 2). This method can also be applied to formalin-fixed single-cell suspensions (e.g., cell culture cells, ascitic or pleuritic effusions). It can also be used on formalin-fixed, paraffin-embedded specimens of various other tissue types (e.g., ovary, endometrium, vagina, kidney, stomach and esophagus).

The use of paraffin-embedded material for flow cytometric DNA analysis has been severely criticized in the past. Indeed, paraffin-embedded tissues are much more susceptible to methodological problems than fresh tissues [6]. Paraffin-embedded tissues often produce a considerable percentage of debris which may affect S-phase calculation [7]. This problem can be easily controlled by good laboratory practice as well as adequate computer-based data evaluation. Also, some fixation methods, other than formalin fixation, yield distorted results if the standard tissue extraction protocol is used [7]. If the protocol cannot be modified accordingly, one should refrain from using such tissue samples. Formalin fixation, which give satisfactory results, is now being used by the majority of pathology departments. The use of external diploid standards can also not be recommended: artifacts caused by fixation or storage might cause variations in binding of the fluorescent dye [1]. However, in any given formalin-fixed paraffin section, there will always be a percentage of benign cells which constitute an adequate diploid standard population [8]. Close cooperation with the pathologist is certainly required in order to assure a consistent high percentage of tumor cells in the samples [9]. Another analytical problem may be the width of the peaks of the DNA-histogram: paraffin-embedded material tends to produce higher coefficients of variation (cv) than fresh tissue [8]. Thus, some near-diploid peaks might be overlooked.

Despite all these drawbacks, several studies have shown that formalin-fixed, paraffin-embedded material and fresh tumor tissue indeed yield comparable DNA-staining results [1,10]. Fresh tissue — if available — is certainly the best choice and to be preferred for flow cytometric DNA analysis because of less time consuming extraction techniques and its better DNA resolution. However, assuming optimized experimental control, paraffin-embedded tissue sections constitute a valuable tool for retrospective analyses on archived material. Moreover, with only one 50 µm thick section required for flow cytometric analysis, they also offer an alternative for analysis of very small tumors. In such cases where only a limited amount of tissue is available for the yet increasing number of prognostic factors to be determined, image analysis has been advocated for assessment of proliferation [11]. The use of paraffin-embedded tissue also allows direct histological control by means of a section cut parallel to that for flow cytometry. In addition, immunohistochemical staining of parallel cut sections, for other proliferation markers or additional antigens, becomes feasible. However, even more so than in fresh tissue, evaluation software that can be geared towards the particular features of paraffin-embedded tissue is essential.

Flow cytometric DNA analysis of fresh tumor cells

Numerous techniques have been proposed in the past for flow cytometric DNA analysis of cells derived from fresh tumor tissue. In 1983 Vindelov et al. introduced

68

a technique for the proteolytic release of nuclei from solid tissues [12]. Degradation of extracellular matrix (stroma) and of cell membranes is achieved by trypsinization (and Nonidet P 40), thus allowing access of DNA stains into cells, e.g., propidium iodide. Other researchers used frozen tumor samples that were sent to their laboratory for routine steroid hormone receptor measurements [13]. The protocol specifies that frozen tumor specimens be pulverized, debris removed on a sucrose gradient and the cell pellets resuspended in FCS-containing medium. Cell membranes are further

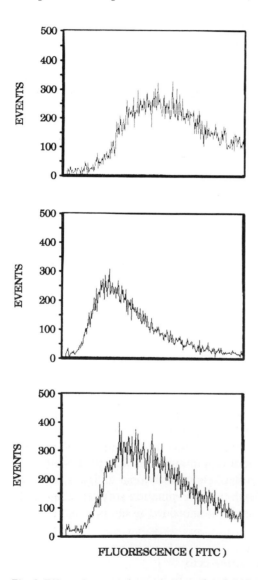

FLUORESCENCE (FITC)

Fig. 3. Effect of enzymatic treatment on the uPA binding capacity of the uPA receptor on unfixed U 937 cells. Control experiment without any collagenase D treatment (upper plot). After treatment with 0.025% collagenase D only 61% of the functional capacity remains (middle), compared to 85% after treatment with 0.005% collagenase D (lower plot).

disintegrated by transfer of the pellet into a subsequent hypotonic sodium citrate buffer, thus leading to a nuclei suspension that could be stained by propidium iodide.

One of the greatest advantages of using fresh tumor tissue for flow cytometric analysis of nuclei is the possibility of determining various other parameters such as histomorphological parameters and antigen expression, thus providing further information about the tumor. This advantage is lost if the techniques employed for tissue degradation completely destroy or severely impair surface or cytoplasmic antigens. For example, inadequate enzymatic treatment with too strong a concentration of collagenase D destroys the function of the uPA receptor (urokinase-type plasminogen activator receptor), while mild treatment with collagenase D leaves it about 85% intact (Fig. 3). Therefore, care must be exercised in the choice of tissue disintegration method. Depending on the dissection techniques applied, not all enzymatically dissected fresh tumor specimens will give uniform quantitative or qualitative results concerning antigen presence (e.g., for the uPA receptor).

Several studies have shown that staining of tumor cells with monoclonal antibodies may facilitate tumor cell identification [14,15]. For instance, antibodies against cytokeratins recognize intermediate filaments of epithelial cells and do not stain inflammatory cells [16]. Thus, by flow cytometric gating on cytokeratin-positive cells, one can exclude tumor-infiltrating, accessory cells from the DNA-histogram that are otherwise present within the diploid population. Assuming adequate tissue disintegration, this technique involving antibodies will help to reduce debris, provide more accurate information about ploidy and S-phase and could reveal near-diploid tumor cell populations that would have been missed otherwise [14]. Since there is no tumor-cell specific antibody, this technique does not exclude benign epithelial cells from the DNA histogram.

We have recently applied the "MACS" technique (magnetic cell separation system) [17] to isolate single competent and unlabeled tumor cells from fresh tumor tissue by a "negative tumor cell selection" procedure [18]. For this purpose fresh carcinoma tissue is subjected to mechanical disintegration immediately after surgery in order to obtain a single-cell suspension. Collagenase D (Boehringer-Mannheim, Germany) at a concentration of 0.001% is added to disintegrate any remaining cell aggregation. Cells other than tumor cells are then labeled with a set of monoclonal antibodies directed to cell surface antigens: CD3 (T-cells), CD14 (macrophages, monocytes), CD15 (granulocytes), CD45R (B-cells) and 5B5 (fibroblasts). Subsequently, anti-isotype antibodies, coupled to ferrit microbeads (Miltenyi, Bergisch-Gladbach, Germany), are incubated with the cell suspension. Those cells reacting with the microbeads are retained on a steel wool matrix under the influence of a magnetic field. Tumor cells that were still unlabeled and thus not reacting with the microbeads, were released from the steel wool matrix by a simple washing step (Fig. 4). This procedure, which takes about 1 h, enables fast and simple isolation of single, living and competent tumor cells from fresh tumor tissue as well as from ascitic or pleuritic effusions. Tumor cell purity is about 93%; after repetition of the same procedure about 97% purity can be obtained (at a recovery rate of 75 or 50%, respectively) (Fig. 5). The unlabeled, intact tumor cells can then be further analyzed by flow

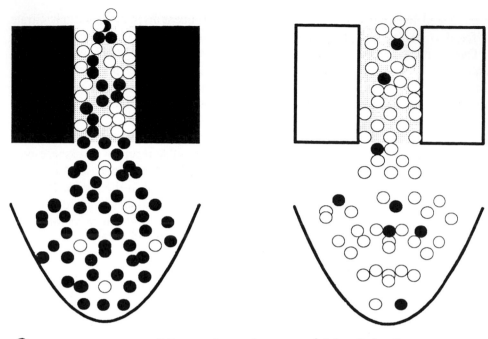

● = **tumor cells & benign epithelial cells**
○ = **cells labelled with monoclonal antibodies & ferrit microbeads**

Fig. 4. Schematic outline of isolation of intact tumor cells from fresh tumor tissue applying the MACS-system: Benign cells (e.g., lymphocytes, leukocytes, fibroblasts), labeled with surface specific antibodies and then reacted with ferrit microbeads, are retained on the steel wool matrix in a strong magnetic field. Nonlabeled cells pass through, resulting in a nearly homogeneous fraction of single tumor cells (left). The retained cells are detached simply by removing the magnetic field and washing with PBS (right).

cytometry and/or confocal laser scanning microscopy for the presence of various surface antigens, including receptors for proteases or growth factors. In addition, after fixation and detergent treatment, nuclear DNA staining for ploidy and S-phase determination becomes possible as well as simultaneous labeling of intracellular antigens.

Computer-based data analysis

Flow cytometric DNA analysis has been performed for more than a decade, but until recently researchers have been reluctant to perform S-phase analyses as a measure of cell proliferation [19]. Several technical aspects have to be considered. Excess debris in disintegrated paraffin-embedded tissue could lead to falsely high S-phase values. Also, tumors with more than one stem cell population (i.e., aneuploid or multiploid

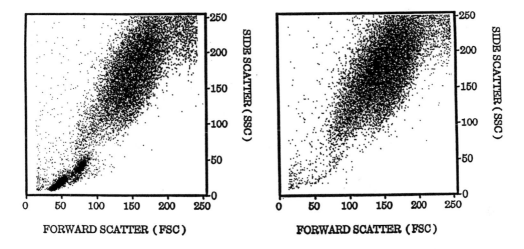

Fig. 5. Enrichment of ovarian carcinoma cells from a suspension of lymphocytes and monocytes. Scatter analysis of the enrichment of tumor cells from 60% (left) before to 97% (right) after MACS. Content of the contaminating lymphocytes and monocytes was reduced from 40% (left) before to 3% (right) after MACS.

tumors) tend to cause analytical problems due to overlapping cell cycles [20]. Last but not least, aggregated cells could cause falsely elevated peaks or mimic additional populations. Solutions to these problems have been proposed utilizing either hardware or software modifications. Hardware modifications, like pulse-processors for doublet removal, attempt to reduce debris or aggregation during flow cytometric data acquisition thus altering the original list mode data. There is a tendency for producers of hardware modifications to oversimplify reality. For example, pulse processors assume that all cells or nuclei are spherical and, if aggregated, present in a straight line. Under these premises the effectiveness of such hardware modifications may be limited. Moreover, such modifications leave the list mode data permanently biased. Further software compensation should therefore not be applied to such modified data [21]. Yet another permanent bias is caused by adding internal standards directly to the tumor sample. Aggregated standard cells (nuclei) would interfere with tumor cell cycle populations. In contrast, merging of separately measured standard-cell (nuclei) samples is still possible at a later stage (e.g., with software).

Recently, several software packages have been presented which attempt exact mathematical modeling of the "real" DNA content while accounting for methodological problems like debris or aggregation at the same time [22,23]. Such software modifications will certainly allow more reliable and reproducible flow cytometric DNA analysis, particularly in S-phase estimation. Correlation of flow cytometrically obtained S-phase results with those of other proliferation assays, like the thymidine labeling index, can also be enhanced by appropriate software modeling [24]. In addition to containing mathematically precise algorithms, adequate software programs should be able to adjust to different tissue types (e.g., paraffin-embedded tissue and fresh tumor) and their unique properties. Weaver et al. were able to show good correlation between S-phase results from paraffin-embedded tissue and fresh tissue after applying appropriate mathematical models [19]. Moreover, "optimal" DNA-evaluation software should be compatible with the most common software platforms to allow convenient data presentation without information loss.

In view of these advances in computer-based analysis of flow cytometrically obtained DNA data, one is reluctant to modify hardware data in any way that could lead to unforeseen and permanent biases. Nevertheless, even very sophisticated software does not produce DNA data that can be transferred to the clinical setting immediately. For example, "cut-off" values for S-phase levels have to be established in a consistent patient cohort accounting for the underlying method of tissue extraction as well as the histological tumor type [25]. Applying S-phase "cut-offs" from the literature without critical evaluation is not appropriate and will impair clinical usefulness. For example, benign and malignant cells can hardly be distinguished in paraffin-embedded diploid tumor samples. Therefore the possibility of different "cut-offs" for aneuploid and diploid tumors must be considered [19].

Summary

Numerous studies have shown that proliferation parameters as assessed by thymidine labeling index, image cytophotometry, DNA flow cytometry or immunohistochemical staining of proliferation-associated antigens are independent and significant prognostic factors in primary breast cancer [2,10,26,27]. In addition, proliferation parameters can be used as predictors of treatment response to chemotherapy or endocrine therapy [28,29]. However, in view of the increasing number of techniques to quantify proliferative potential of tumor cells, more data will certainly be needed before a consensus on the following issues can be reached:

— Do proliferation markers in general have a prognostic impact in primary breast cancer that is greater than that of the "classical" prognostic factors, or will they be more important as part of a combined prognostic index?

— Is any single proliferation marker particularly suited for clinical application in primary breast cancer due either to its outstanding prognostic impact or to its standardized determination method?

While any definite recommendations may be premature, rather time consuming techniques like image cytometry or thymidine labeling index might not be useful on a routine basis. However, they can provide important additional information in certain cases [30]. When choosing a particular proliferation marker for everyday clinical application, technical considerations should play an important role. For none of these markers has a standardized determination method been successfully established, and thus clinical usefulness is still limited. It is therefore preferable to be able to perform laboratory assays and data evaluation at the same institute in order to assure optimal quality control.

Flow cytometric DNA analysis, both on fixed paraffin-embedded tissue and on fresh tumor tissue, offers rapid and easy assessment of the proliferative activity as well as the DNA content of a tumor. It constitutes a powerful tool for retrospective and prospective studies. Since evaluation steps can be automated, analytic subjectivity is minimized. Considerable statistical power is added by the large number of cells that can be screened per sample. Correlation of the results with older techniques that might be considered "gold standard proliferation assays" (i.e., thymidine labeling index or image cytometry) has been shown by many researchers [24,30,31]. In fresh tumor samples, simultaneous immunocytochemical staining with monoclonal antibodies can add valuable information about the tumor. The prognostic impact of flow cytometrically determined S-phase and ploidy in primary breast cancer has been clearly demonstrated [25,26,32,33]. However — as outlined in this article — tissue extraction and computer-based data analysis are not free of technical pitfalls. Therefore, international standardization of these methods is urgently needed before results of flow cytometric DNA analysis can be transferred to routine clinical management.

Acknowledgements

This work was supported by the "Klinische Forschergruppe" GR 280/4-1 der Deutschen Forschungsgemeinschaft and a grant of the Wilhelm Sander Stiftung.

References

1. Camplejohn RS. In: Herrington, McGee (eds) Diagnostic Molecular Pathology: A Practical Approach. Oxford: IRL Press, 1992.
2. Tubiana M, Pejovie MH, Chavaudra N, Contesso G, Malaise EP. Int J Cancer 1984;33:441–445.
3. Fallenius AG, Franzen SA, Auer GU. Cancer 1988; 62:521–530.
4. Hedley DW, Friedlander ML, Taylor IW, Rugg CA. Musgrove EA. J Histochem Cytochem 1983; 31(11):1333–1335.
5. Harbeck N, Moniwa N, Busch E, Schmitt M, Jänicke F, Fellbaum C, Höfler H, Graeff H. Gynäkol Rundsch 1991;31(suppl 2):299–302.
6. Merkel DE, McGuire WL. Cancer 1990;65:1194–1205.
7. Hedley DW. Cytometry 1989;10:229–241.
8. Hedley DW, Friedlaender ML, Taylor IW. Cytometry 1985;6:327–333.
9. Kute TE, Gregory B, Galleshaw J, Hopkins M, Buss D, Case D. Cytometry 1988;9:494–498.
10. Kallioniemi OP. Cytometry 1988;9:164–169.
11. Ghali VS, Liau S, Teplitz C, Prudente R. Cancer 1992;70:2668–2672.
12. Vindelov LL, Christensen IJ, Nissen NI. Cytometry 1983;3(5):323–327.
13. Dressler LG, Seamer LC, Owens MA, Clark GM, McGuire WL. Cancer 1988;61:420–427.
14. Ferrero M, Spyratos F, Le Doussal V, Desplaces A, Rouesse J. Cytometry 1990;11:716–724.
15. Van der Linden JC, Herman CJ, Boenders JGC, van de Sandt MM, Lindeman J. Cytometry 1992; 13:163–168.
16. Moll R, Franke WW, Schiller DL, Geiger B, Krepler R. Cell 1983;31:11–24.
17. Miltenyi S, Müller W, Weichel W, Radbruch A. Cytometry 1990;11:231–238.
18. Harbeck N, Yamamoto N, Schüren E, Schmitt M, Höfler H, Graeff H. Eur J Cancer 1993;29A (suppl 6):S64.

19. Weaver DL, Bagwell CB, Hitchcox SA, Whetstone SD, Baker DR, Herbert DJ, Jones MA. American J Clin Pathol 1990;94(5):576–584.
20. Dressler LG, Seamer L, Owens MA, Clark GM, McGuire WL. Cancer Res 1987;47:5294–5302.
21. Bagwell CB. DNA data analysis. Workshop at the 5th annual flow cytometry forum, San Antonio, TX, 1992.
22. Bagwell CB, Mayo SW, Whetstone SD, Hitchcox SA, Baker DR, Herbert DJ, Weaver DL, Jones MA, Lovett EJ. Cytometry 1991;12:107–118.
23. Rabinovitch PS. Multicycle. Phoenix Flow Systems, San Diego, CA, 1988.
24. Haag D, Feichter G, Goerttler K, Kaufmann M. Cytometry 1987;8:377–385.
25. Sigurdsson H, Baldetrop B, Borg A, Dalberg M, Fernoe M, Killander D, Olsson H. N Engl J Med 1990;322:1045–1053.
26. Clark GM, Dressler LG, Owens MA, Pounds G, Oldaker T, McGuire WL. N Engl J Med 1989;320: 627–633.
27. Weikel W, Beck T, Mitze M, Knapstein PG. Breast Cancer Res Treat 1991;18(3):149–154.
28. Osborne CK. J Natl Cancer Inst 1989;81(18):1344–1345.
29. Briffod M, Spyratos F, Hacene K, Tubiana-Hulin M, Pallud C, Gilles F, Rouesse J. Cytometry 1992; 13:250–258.
30. Baldetorp B, Ferno M, Fallenius A, Fallenius-Vecchi G, Idvall I, Olsson H, Sigurdsson H, Akerman M, Killander D. Cytometry 1992;13:577–585.
31. McDivitt RW, Stone KR, Craig RB, Palmer JG, Meyer JS, Bauer WC. Cancer 1986;57:269–276.
32. Toikkanen S, Joensuu H, Klemi P. Br J Cancer 1989;60:693–700.
33. Harbeck N, Schüren E, Pache L, Schmitt M, Jänicke F, Dettmar P, Höfler H, Graeff H. Breast Cancer Res Treat 1992;23:149.

Tumor invasion and metastasis

Structure and function of matrix metalloproteinases and their significance in breast cancer

H. Tschesche[1], J. Bläser[1], T. Kleine[1], W. Bode[2], P. Reinemer[2], F. Grams[2], S. Schnierer[1], C. Thomssen[3], M. Schmitt[3], L. Pache[3], F. Jänicke[3] and H. Graeff[3]

[1]*Lehrstuhl für Biochemie, Fakultät für Chemie, Universität Bielefeld, 33615 Bielefeld;* [2]*Max-Planck-Institut für Biochemie, 82152 Martinsried; and* [3]*Frauenklinik der Technischen Universität, Klinikum rechts der Isar, 81675 München, Germany*

Tumour growth involves special tumour cell interaction with the extracellular matrix and in particular with the basement membranes. Tumour spreading and the dissemination of tumour cells from their site of origin to distant sites in the body, referred to as metastasis, require the proteolytic degradation of the surrounding connective tissues [1–3]. This final process occurs as the outcome of a multifactorial process involving cellular transformations with modifications in gene expression [4]. As a result, we may observe uncontrolled proliferation of tumour cells, facilitated attachment by adhesion molecules to host cellular or extracellular matrix elements, increased proteolysis of host barriers for spreading and invasion and tumour cell locomotion and colony formation.

Degradation of extracellular matrix

The unspecific proteolytic breakdown of extracellular matrix structural macromolecules can be accomplished in vitro by enzymes of several different classes of proteinases, e.g., serine proteinases, such as plasmin or leucocyte proteinases, cysteine proteinases, such as cathepsins B, H and L, or aspartyl proteinases, such as cathepsin D.

Specific cleavage of extracellular matrix components is achieved by the matrix metalloproteinases (MMPs). The nine members of this MMP gene family established so far exhibit proteolytic activity against most, if not all extracellular matrix macromolecules (Table 1). The two interstitial collagenases (fibroblast collagenase, MMP-1, and leucocyte collagenase, MMP-8) are homologous proteins encoded by two different genes. Both cleave types I, II and III collagen into specific three-quarter and one-quarter fragments, whereby MMP-1 shows a preference for type III and MMP-8 for type I collagen. This specificity is not shared by other members of the MMP family. While the stromelysins 1 and 2 and matrilysin are less specific in substrate degradation the gelatinases (72 kDa gelatinase, MMP-2 and 92 kDa gelatinase, MMP-9) specifically cleave basement membrane collagens types IV and

Table 1. Matrix metalloproteinase family

Enzyme	MMP#	M_r	Extracellular matrix substrates	Cellular and Source
Fibroblast-type collagenase	MMP-1	57,000/ 52,000	Collagen I, II, III (III>>1) VII, VIII, X; gelatin; PG core protein	Fibroblasts, keratinocytes, endothelial cells, monocytes, macrophages, chondrocytes and osteoblasts [3]
PMN-type collagenase	MMP-8	75,000	Same as fibroblast-type collagenase (I>>III)	Leucocytes [3]
Stromelysin-1	MMP-3	60,000/ 55,000	PG core protein; fibronectin; laminin; collagen IV, V, IX, X; elastin; proCL	Fibroblasts, endothelial cells [3]
Stromelysin-2	MMP-10	60,000/ 55,000	Same as SL-1	Fibroblasts, keratinocytes [3]
Stromelysin-3	MMP-11	n.d.	n.d	Fibroblasts, carcinoma cells[a]

n.d. = not determined.
MMP numbering according to Nagase H, Barrett AJ, Woessner Jr. JF et al. Matrix Spec. Suppl 1. 1992;421–424.
[a]Bassett P, Wolf C, Chambon P. Breast Cancer Res Treat 1993;24:185–193.

(continued)

Table 1. Matrix Metalloproteinase Family

Enzyme	MMP#	M_r	Extracellular matrix substrates	Cellular and Source
Macrophage metallo-elastase	?	53,000	Elastin	Macrophages[b]
M_r 72K gelatinase/ type IV collagenase	MMP-2	72,000	Gelatin; collagen I, V, VII, X, XI; elastin; fibronectin; PG core protein	Fibroblasts, keratinocytes, chondrocytes, endothelial cells, monocytes, osteoblasts, malignant transformed cells [3]
M_r 92K gelatinase/ type IV collagenase	MMP-9	92,000	Gelatin; collagen IV, V; elastin, PG core protein	Leucocytes, keratinocytes, monocytes and macrophages, malignant transformed cells [3]
Matrilysin (PUMP-1)	MMP-7	28,000	Fibronectin, laminin, collagen IV, gelatin proCL, PG core protein	Cancer cells[c,d], endometrial cells[e], monocytes[f]

n.d. = not determined.

MMP numbering according to Nagase H, Barrett AJ, Woessner Jr. JF et al. Matrix Spec. Supp 1: 1992;421-424.

[b]Shapiro SD, Kobayashi DK, Grey TJ. J Biol Chem 1993;268:23824–23829.

[c]Yoshimoto M, Itoh F, Yamamoto H, Hinoda Y, Imai K, Yachi A. Int J Cancer 1993;54:614–618.

[d]Powell WC, Knox JD, Navre M, Grogan TM, Kittelson J, Nagle RB, Bowden GT. Cancer Res 1993;53:417–422.

[e]Rodgers WH, Osteen KG, Matrisian LM, Navre M, Guidice LC, Gontein F. Am J Obstet Gynecol 1993;168:253–260.

[f]Busiek DF, Ross FP, McDonnell S, Murphy G, Matrisian LM, Welgus HG. J Biol Chem 1992;267:9087–9092.

80

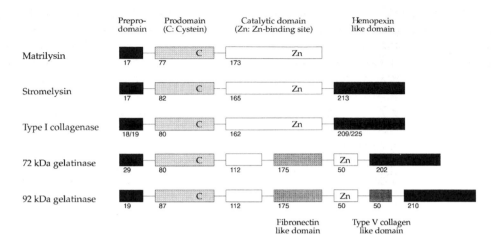

Fig. 1. Domain structure of MMPs.

V into three-quarter and one-quarter fragments and are therefore also designated type IV collagenases. Stromelysin 3, first detected in breast cancer tissue [5,6], forms another group within the MMPs [7] with as yet unknown specificity.

The enzymes are released from various types of normal cells (Table 1) as proenzymes, i.e., inactive precursor forms also designated latent enzymes. In general their secretion from fibroblast type cells occurs as a result of biosynthetic activity of the cells without prior storage. However, in leucocytes the leucocyte collagenase MMP-8 and the 92 kDa gelatinase/type IV collagenase MMP-9 are stored in specific granules and C-particles after biosynthesis during cellular differentiation. Their release requires uptake of a special signal to the phagocytic cell. The unusually high carbohydrate content of the leucocyte collagenase, amounting to one third of the total molecular weight of the activated enzyme [8], could be of relevance for the storage properties of the enzyme. In comparison, the corresponding fibroblast collagenase bears only a small carbohydrate moiety [9].

Domain assembly of the MMPs

A multidomain structure of the MMPs is a general feature of all nine or more members of this protein family (Fig. 1). The hydrophobic signal peptide sequence comprising 17–19 residues, as deduced from the cDNA sequence, is not present in the secreted forms of the MMPs. The secretory precursor form starts with the N-terminal propeptide domain of about 80 residues, which provides latency of the enzyme. The following catalytic domain contains the zinc binding site and the reactive residues providing the catalytic activity. A hemopexin-like C-terminal domain is linked by a 5–50 residues proline-rich hinge region to the catalytic domain, which was shown to play a role in encoding substrate specificity for the interstitial collagenases [10–12]. Only matrilysin (i.e., PUMP-1) lacks the hemopexin-like domain and is, therefore, considerably smaller than all the other enzymes. The two type IV collagenases/gelatinases (MMP-2 and MMP-9) contain an additional insert in the catalytic domain built of three tandem repeats of a fibronectin-like structure, which mediate gelatin-binding properties to these enzymes. In addition, the 92 kDa collagenase (MMP-9) contains a type V collagen-like insert (Fig. 1).

Besides one single unpaired Cys residue in the propeptide domain, which is believed to occupy the fourth ligand site of the catalytic zinc in the inactive (latent) precursor form (see below), only two further Cys residues are present in the hemopexin-like domain forming a single disulfide bridge with the exception of MMP-9, which has three Cys residues in its hemopexin-like domain.

Activation of latent MMPs

The MMPs are secreted as latent precursor forms and require extracellular activation. Though various proteolytic enzymes, e.g., trypsin, chymotrypsin, cathepsin G and kallikrein amongst others, are rapid activators, by cleaving the propeptide domain about 80 residues from the N-terminus, the physiological pathway of activation is not yet clear. In vivo activation of interstitial fibroblast procollagenase was ascribed to the action of plasmin, which could be released from the activation cascade of plasminogen by plasminogen activator. However, neither the 72 kDa nor the 92 kDa procollagenase/gelatinase, nor the interstitial leucocyte procollagenase are subject to activation by plasmin [3,8]. However, the 72 kDa gelatinase (type IV collagenase, MMP-2) can be activated by uPA [13,14].

All enzymes are indeed rapidly activated by the proteolytic action of stromelysin or by autocatalytic processing [3]. Besides proteolytic activation various nonproteolytic agents such as SDS, HOCl, Hg-derivatives, thiol reagents, oxidants and chaotropic agents, such as KI or KSCN, lead to activation. This behaviour is explained by the now generally accepted cysteine switch hypothesis [15].

The cysteine switch activation mechanism

The single unpaired Cys of the strongly conserved PRCGVPD sequence motif within the propeptide domain is assumed to provide the fourth coordination ligand of the active site zinc. Activation requires replacement of the coordinating Cys moiety by a water molecule. This opens the reactive site and provides access of substrates to the established catalytic machinery (Fig. 2). The opening of the switch is either achieved by disruption of the coordinate Cys zinc bond followed by several autolytic cleavages to generate the active and finally processed enzyme from or by successive proteolytic cleavages by those proteinases, which are able to excise portions of the propeptide domain (Fig. 2). The successive proteolytic activations have been studied in detail, e.g., in leucocyte procollagenase activation [16]. It is interesting to note that the

82

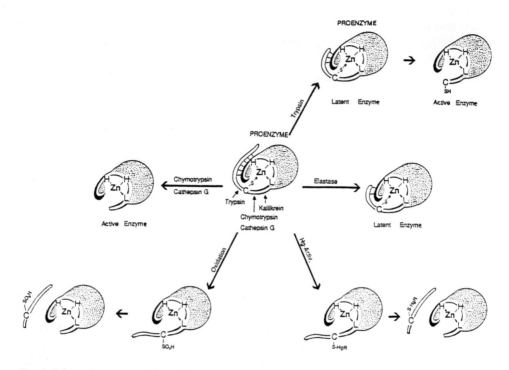

Fig. 2. Schematic representation of the different modes of activation of leucocyte (PMNL)-procollagenase, e.g. by proteolytic cleavage of the propeptide domain by various serine proteinase or by mercury compounds or oxidising agents representing the cysteine switch mechanism [15].

inflammatory enzyme from human granulocytes is not activated by human leucocyte elastase, the predominant proteolytic enzyme of this cell. It is, however, rapidly activated by the other chymotrypsin-like proteinase, cathepsin G (Fig. 3).

The three-dimensional structure of MMPs

The recombinant catalytic domain of human leucocyte interstitial collagenase has been expressed in *Escherichia coli* (*E. coli*) [11]. It was crystallised and the crystal structure determined at 2.0 Å resolution [17]. The spherical molecule contains a shallow active site cleft separating a smaller C-terminal subdomain from a bigger N-terminal main domain, which is composed of a central, highly twisted five-stranded β-sheet, flanked by an S-shaped double loop and two other bridging loops on its convex side and by two long α-helices on its concave side. The catalytic zinc ion is located at the bottom of the active site cleft and is coordinated by the $N_{\epsilon 2}$-atoms of the three His within the His[197]-Glu[198]-X-X-His[201]-X-X-Gly[204]-X-X-His[207] zinc binding consensus sequence. The active site helix contains His[197], Glu[198] and His[201] (leucocyte collagenase numbering [16]) and extends to Gly[204], where the polypeptide chain turns away from the helix axis towards the third zinc ligand, His[207].

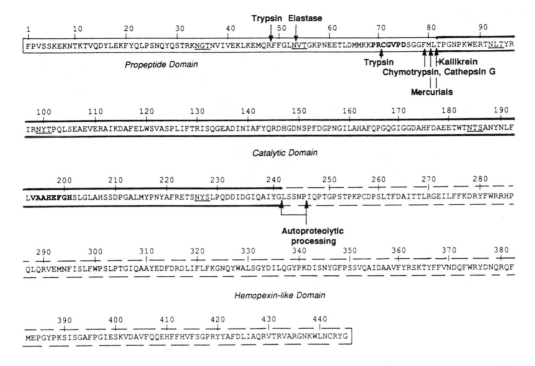

Fig. 3. Human leucocyte (PMNL) procollagenase amino acid sequence and domain structure with proteolytic and autoproteolytic processing sites [8,12,16].

The side chain of Met[215] in the right-hand loop from Ala[213]-Leu[214]-Met[215]-Tyr[216] designated the "Met-turn" provides a hydrophobic environment for the three zinc-ligating His residues. This element is characteristic and common to the subfamily of enzymes designated as "metzincins" [18].

The peptide chain is kinked at Pro[217] forming the outer wall of the putative S 1′-substrate pocket. The chain then forms the C-terminal helix on the bottom of the molecule containing the strictly conserved residues Asp[232] and Asp[233]. The latter residue, completely buried, is essential for the catalytic activity [19]. It forms two hydrogen bonds to the amide nitrogens of the Met-turn residues Leu[214] and Met[215] stabilising the active site basement (Fig. 4).

Besides the "catalytic" zinc ion, a second "structural" zinc ion is sandwiched between the surface S-shaped double loop Arg[145]-Leu[160] and the surface of the β-sheet. It is tetrahedrally coordinated by His[147], Asp[149], His[162] and His[175] while a structural calcium ion is octahedrally coordinated by Asp[137], Gly[155], Asn[157], Asp[177] and Glu[180]. A second structural calcium ion is located on the convex side of the β-sheet at the end of the glycine-rich open loop. It is also octahedrally coordinated by Asp[137], Gly[169], Gly[171], Asp[173] and by two water molecules (Fig. 4).

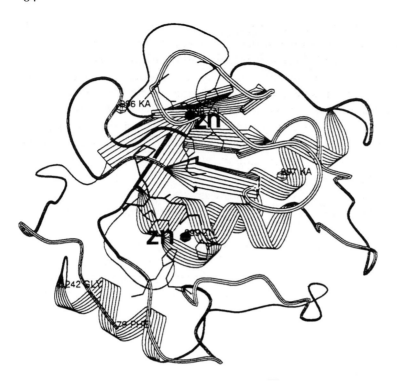

Fig. 4. Structure of the Phe79-Gly242 catalytic domain of human leucocyte (PMNL) collagenase shown as a ribbon model. An upper (N-terminal) and a lower (C-terminal) domain are separated by a moderately deep active site cleft. The upper domain harbours two zinc ions (Zn998 on the top, coordinated by three His and one Asp) and the "catalytic" zinc (Zn999 in the middle, coordinated by three His and the hydroxamic moiety of an inhibitor (thick line of Pro-Leu-Gly-NHOH) and two calcium ions (Ka996 and Ka997, displayed as spheres without coordinating ligands). Taken from Reinemer P, Grams F, Huber R, Kleine T, Schnierer S, Pieper M, Tschesche H, Bode W. FEBS Letts 1994;338:227–233.

Substrate recognition and specificity

Cleavage of native triple-helical collagen into three-quarter and one-quarter fragments requires proteolysis of three peptide bonds in three separate super-coiled peptide strands. This specific "triple-helicase" activity of interstitial fibroblast and leucocyte collagenase requires the C-terminal hemopexin-like domains [10–12]. The single catalytic domain exhibits proteolytic activity against various peptide substrates (Dieckmann and Tschesche, in preparation), but not against native type I collagen [10–12]. We assume that the repetitive Pro-X-Gly segment of one substrate strand of the three triple-helical collagen structure could probably orient itself against the nonprimed collagenase subsites in such a manner that the glycyl carbonyl group approaches the catalytic zinc. In such a collagen-collagenase complex the substrate P1′ proline side chain would not fill the S1′ subsite pocket adequately, and the collagen triple helix 15 Å in diameter would be too bulky to fit through the opening at the S2'-S3' subsites. Therefore, the proper adjustment of the scissible peptide chain to the enzyme's binding subsites necessarily seems to require at least partial unfolding of the collagen triple helical structure around the cleavage site.

The susceptibility to cleavage by interstitial collagenases requires hydrophobic P1' residues, such as leucine or isoleucine and the absence of proline residues in the primed sites of triple helical type I collagen molecules, as revealed by mutagenesis experiments and expression of murine α1(I) collagen [20]. This resistance to cleavage of the mutant proteins is not a property of the isolated triple helix itself, since the noncleavable mutant collagens do not exhibit enhanced melting temperatures. Obviously, this property is the result of the collagen-collagenase complex. The residues flanking the cleavage site of the scissile collagen peptide bond might be released from their triple-helical interactions due to energetically favourably interaction with the enzyme subsites.

In collagen substrate complexes with the full length collagenase, the rod like triple helix might be clamped between the catalytic and the hemopexin-like domain. Such a "clamping mechanism", allosterically feasible by the flexible hinge region between both domains, could sandwich the flanking regions of the scissile peptide bond and could structurally and temporarily fix the triple helix in the destabilised form to facilitate cleavage. Rotation of the collagen substrate strand around its helix axis would allow the "triple helicase" activity typical for the interstitial collagenases.

MMPs in tumour tissue

Tumour tissue is generally found to exhibit various proteolytic activities. These "tumour cell associated" protease activities are either membrane bound, intracellular or secretory. The MMPs are secretory enzymes, but have also been found associated with the cell surface. In human breast adenocarcinoma cells, MDA-MB 231 and MCF-7, a membrane associated putative receptor for the 72 kDa type IV collagenase was demonstrated [21] as was previously evidenced for the urokinase type plasminogen activator. Both the 72 kDa and the 92 kDa type IV collagenase have been detected in various tumour tissues, e.g., immunoblastic lymphomas [22], mammary carcinomas [23–25], melanoma cells [26], lung fibroblasts [27], pancreatic tumour cells [28], squamous and basal cell carcinomas [29,30], and ovarian tumour cells [31], and in the culture medium of various transformed cell lines, e.g., gastric and colorectal carcinomas [32], endometrial carcinomas [33], osteosarcoma, fibrosarcoma and histiocytoma, [34] to mention but a few.

It was therefore tempting to determine the content of MMPs in human breast cancer tissue and to evaluate a putative correlation with the probability of increased risk of relapse and death. The tissues were extracted as published previously [35]. Determinations of MMP-8, MMP-9 and TIMP-1 were carried out by highly specific enzyme-linked immunosorbent assay (ELISA) techniques as described earlier [36,37]. The other factors of the general disease status such as axillary node status, menopausal status, hormone receptor status and tumour diameter were also

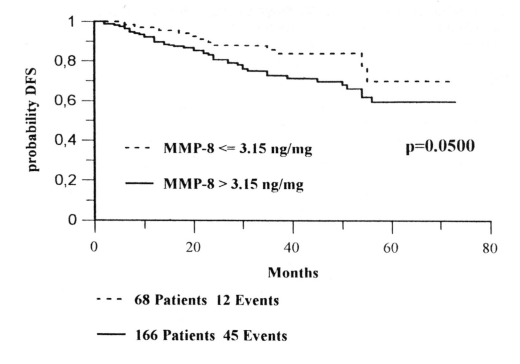

Fig. 5. Disease-free survival (DFS) as a function of leucocyte (PMNL) collagenase (MMP-8) in breast cancer patients (n = 234). Univariate analysis. Patients with an MMP-8 content higher than 3.15 ng/mg protein have an increased risk to relapse.

determined, as were uPA and PAI-1 content. All parameters were used for statistical analyses. The disease-free survival was analysed according to the Cox model [38] and the curves were calculated in accordance with the Kaplan-Meier method [39].

The study was performed with 234 patients, 97 of whom were node-negative. The median follow up was 35 months, ranging from 1 to 75 months. The median age was 55.4 years, ranging from 27.3 to 88.6 years.

The univariate analysis of node-negative patients showed that neither MMP-8 nor MMP-9 nor TIMP-1 were of significant prognostic value for disease-free survival. However, in the univariate analysis covering all patients, the leucocyte collagenase, MMP-8, had some statistical significance with MMP-8 levels higher than 3.15 ng/mg protein (p = 0.0500) (Fig. 5). This statistical significance might indicate the inflammatory response.

Acknowledgements

This work was supported by the Deutsche Forschungsgemeinschaft, Special Research Programme SFB 223, project B01 and the Klinische Forschergruppe GR280/4−1 and the Fonds der Chemischen Industrie (H.T.). The authors wish to thank Mrs V. Süwer,

Mrs E. Sedlaczek and Mrs B. Jaud-Münch for skilful technical assistance and Mrs G. Delany for linguistic advice.

References

1. Liotta LA, Rao CN, Wewer UM. Ann Rev Biochem 1986;55:1037–1057.
2. Stetler-Stevenson WG. Cancer Metastasis Rev 1990;9:289–303.
3. Birkedal-Hansen H, Moore WGI, Bodden MK, Windsor LJ, Birkedal-Hansen B, DeCarlo A, Engler JA. Oral Biol Med 1993;4(2):197–250.
4. Liotta LA, Stetler-Stevenson W, Steeg PS. In: Vincent T, DeVita J, Hellmann S, Rosenberg St. A (eds) Important Advances in Oncology. Philadelphia: Lippincott, 1991;85–100.
5. Basset P, Bellocq JP, Wolf C, Stoll I, Hutin P, Gimacher JM, Podhajcer OC, Chenard MP, Rio MC, Chambon P. Nature 1990;348:699–704.
6. Basset P, Wolf C, Chambon P. Breast Cancer Res Treat 1993;24:185–193.
7. Murphy GJP, Murphy G, Reynolds JJ. FEBS Lett 1991;289:4–76.
8. Tschesche H, Knäuper V, Krämer S, Michaelis J, Oberhoff R, Reinke H. In: Birkedal-Hansen B, Werb Z, Welgus H, Van Wart H (eds) Matrix Metalloproteinases and Inhibitors. Stuttgart-New York: Gustav Fischer Verlag, 1993; Matrix Supplement 1:245–255.
9. Wilhelm SM, Eisen AZ, Teter M, Clark SD, Kronberger A, Goldberg GI. Proc Natl Acad Sci USA 1986;83:3756–3760.
10. Murphy G, Allan JA, Willenbrock F, Cachett MI, O'Connell JP, Docherty AJP. J Biol Chem 1992; 267:9612–9618.
11. Schnierer S, Kleine T, Gote T, Hillemann A, Knäuper V, Tschesche H. Biochem Biophys Res Comm 1993;191:319–326.
12. Knäuper V, Osthues A, DeClerck YA, Langley KE, Bläser J, Tschesche H. Biochem J 1993;291: 847–854.
13. Reith A, Rucklidge GJ, Biochem Biophys Res Comm 1992;186:348–354.
14. Keski-Oja J, Lohi J, Tuuttila A, Tryggvason K, Vartio T. Exp Cell Res 1992;202:471–476.
15. Van Wart H, Birkedal-Hansen H. Proc Natl Acad Sci USA 1990;87:5578–5582.
16. Knäuper V, Krämer S, Reinke H, Tschesche H. Eur J Biochem 1990;189:295–300.
17. Bode W, Reinemer P, Huber R, Kleine T, Schnierer S, Tschesche H. EMBO J 1994;13:1263–1269.
18. Bode W, Gomis-Rüth FX, Stöcker W. FEBS Lett 1993;331:134–140.
19. Hirose T, Patterson C, Pourrmotabled T, Mainardi CL, Hasty KA. Proc Natl Acad Sci USA 1993; 90:2569–2573.
20. Wu H, Byrne MH, Stacey A, Goldring MB, Birchhead JR, Jaenisch R, Krane SM. Proc Natl Acad Sci USA 1990;87:5888–5892.
21. Emonard HP, Remache AG, Noël AC, Grimand JA, Stetler-Stevenson WG, Fordart JM. Cancer Res 1992;52:5845–5848.
22. Kossakowska AE, Urbanski StJ, Huchcroft SA, Edwards DR. Oncology Res 1992;4:233–240.
23. Monteagudo C, Merino MJ, San-Juan J, Liotta LA, Stetler-Stevenson WG. Am J Pathol 1990;136: 585–592.
24. Lyons JG, Birkedal-Hansen B, Moore WGI, O'Grady RLO, Birkedal-Hansen H. Biochemistry 1991; 30:1449–1456.
25. Schmitt M, Jänicke F, Thomssen C, Pache L, Kramer M, Bläser J, Tschesche H, Wilhelm O, Weidle U, Graeff H. In: Preissner KT, Rosenblatt S, Kost C, Wegerhoff J, Mosher D (eds) Biology of Vitronectins and their Receptors. Amsterdam: Elsevier Science Publishers 1993;331–341.
26. Houde H, Masure S, Opdenakker G. Fibrinolysis 1992;6(suppl 2):145–150.
27. Wilhelm MS, Collier IV, Marmer LB, Eisen AZ, Grant GA, Goldberg GI. J Biol Chem 1989;264: 17213–17221.
28. Zucker S, Moll MU, Lysik RM, Di Massimo EI, Stetler-Stevenson WG, Liotta LA, Schwedes JW. Int J Cancer 1990;45:1137–1142.

29. Pyke C, Ralfkiaer E, Huhtala P, Hurskainen T, Danø K, Tryggvason K. Cancer Res 1992;52:1336–1341.
30. Okada Y, Tsuchiya H, Shimitu H, Tomita K, Nakanishi I, Sato H, Seiki M, Yamashita K, Hayakawa T. Biochem Biophys Res Comm 1990;171:610–617.
31. Campo E, Merino MJ, Tavassoli FA, Charonis AS, Stetler-Stevenson WG, Liotta LA. Am J Surg Pathol 1992;16:500–507.
32. Yamagata S, Yoshii Y, Suh JG, Tanaka R, Shimizu S. Cancer Lett 1991;59:51–55.
33. Takemura M, Azumo C, Kimura T, Tokugawa Y, Miki M, Ono M, Saji F, Tanizawa O. Cancer 1992;70:147–151.
34. Sato H, Kida Y, Mai M, Endo Y, Sasaki T, Tanaka J, Seiki M. Oncogene 1992;7:77–83.
35. Jänicke F, Schmitt M, Pache L, Ulm K, Harbeck N, Höfler H, Graeff H. Breast Cancer Res Treat 1993;24:195–208.
36. Bläser J, Triebel S, Kopp C, Tschesche H. Eur J Clin Chem Clin Biochem 1993;31:513–516.
37. Günther M, Haubeck HD, Van de Leuer E, Bläser J, Bender S, Finder DC, Tschesche H, Greiling H, Heinrich PC, Graeve L. Arthritis Rheum 1994 (in press).
38. Cox DR. J R Stat Soc (B) 1972;34:187–220.
39. Kaplan EL, Meier P. J Am Stat Assoc 1958;8:771–783.

Prospects in diagnosis and treatment of breast cancer
M. Schmitt et al., editors

Stromelysin-3 gene expression in breast and other human carcinomas

Jean-Pierre Bellocq[1]*, Marie-Pierre Chenard-Neu[1], Nicolas Rouyer[1], Catherine Wolf[2], Pierre Chambon[2] and Paul Basset[2]

[1]*Service d'Anatomie Pathologique Générale, Centre Hospitalier Régional Universitaire, Hôpital de Hautepierre, 1 avenue Molière, 67098 Strasbourg Cedex; and [2]Laboratoire de Génétique Moléculaire des Eucaryotes du CNRS, Unité 184 de Biologie Moléculaire et de Génie Génétique de l'INSERM. Institut de Chimie Biologique, Faculté de Médecine, 11 rue Humann, 67085 Strasbourg Cedex, France*

Abstract. Stromelysin-3 (ST3) is a new matrix metalloproteinase which exhibits unique structural and functional characteristics. The gene is specifically expressed in fibroblastic cells immediately surrounding cancer cells in most human invasive carcinomas. The ST3 gene is also expressed in noninvasive carcinomas, including some in situ breast carcinomas, where the gene is most frequently expressed in those having the highest probability to become invasive. These results, together with the observation that the ST3 and urokinase genes have very similar expression patterns in breast carcinomas, suggest that ST3 could be involved in cancer progression.

Introduction

Cancer cell invasion at the primary site of carcinomas is thought to involve the co-operation of multiple proteolytic enzymes whose substrates are components of the extracellular matrix [1]. Although many proteinases can cleave matrix molecules, the secreted matrix metalloproteinases (MMPs) and the serine proteinases of the plasmin system are believed to be the most relevant mediators which modify the integrity of the extracellular matrix [2–5].

A strategy based on the differential screening of cDNA libraries has been developed in our laboratory to identify genes that may play a role in breast cancer progression. In order to consider the interactions which occur between neoplastic and stromal cells during tumor progression [6–8], a cDNA library was constructed from surgical specimens rather than from established cancer cell lines. Using this approach we have identified a new MMP that we termed stromelysin-3 (ST3) [9].

Address for correspondence: Jean-Pierre Bellocq MD, Service d'Anatomie Pathologique Générale, Centre Hospitalier Régional Universitaire, Hôpital de Hautepierre, 1 avenue Molière, 67098 Strasbourg Cedex, France.

Table 1. Matrix metalloproteinases and their substrates (modified from Matrisian [5])

Name	Substrates
Collagenases	
Interstitial collagenase	Fibrillar collagens: I, II, III, VII, X
Neutrophil collagenase	Fibrillar collagens: I, II, III
Gelatinases	
Gelatinase A	
(72-kDa type IV collagenase)	Denatured collagens (gelatins), basement membrane collagen
Gelatinase B	IV, collagen V, elastin
(92-kDa type IV collagenase)	
Stromelysin	
Stromelysin-1	Proteoglycans, fibronectin, laminin, collagens III, IV, V, IX,
Stromelysin-2	gelatins
Matrilysin (Pump-1)	Proteoglycans, fibronectin, gelatins, elastin
Other members	
Metalloelastase	elastin, fibronectin
Stromelysin-3	?

ST3 is a new member of the matrix metalloproteinase family

The MMP family now consists of nine distinct members in the human, which can degrade most extracellular matrix components (Table 1). Thus, MMPs are believed to be involved in physiological and pathological processes associated with tissue remodeling and modifications of cell-matrix interactions, such as those occurring during embryogenesis, morphogenesis and cancer progression. The sequence of the ST3 protein, as deduced from that of its cDNA [9], exhibits the general features of previously described MMPs [5], including a typical zinc-binding site in the catalytic domain and the cysteine residue characteristic of the MMP prodomain (Fig. 1). Comparison of ST3 amino acid sequence with those of stromelysin-1 and -2 (ST1 and ST2) indicates that ST1 and ST2 are more similar to each other than to ST3 [9]. Furthermore, ST3 differs from the other MMPs by the presence of 10 amino acid residues at the C-terminal part of the prodomain [9] (Fig. 1). These observations suggested that ST3 may be in fact the first member of a new MMP group [10]. In agreement with this concept, ST3 also differs from the other MMPs by its enzymatic properties. No substrate has yet been demonstrated for ST3, although a C-terminally truncated form of ST3 was found to exhibit weak stromelysin-like activity [11] (Fig. 1). Thus, by partial analogy with interstitial collagenase [12], one hypothesis is that ST3 with an intact C-terminal domain has specific properties for an as yet undefined substrate.

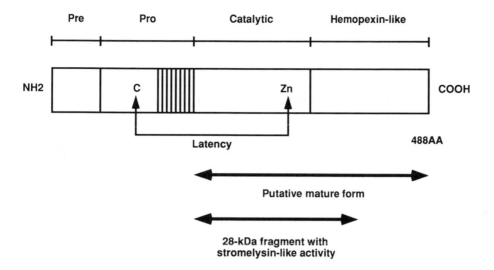

Fig. 1. Protein domains of ST3. The vertical bars in the prodomain indicate the 10 amino-acid residues specific of ST3.

ST3 expression in fibroblastic cells of breast carcinomas

The absence of ST3 gene expression in eight breast cancer cell lines, although present in the 30 breast carcinomas initially tested by Northern blot analysis, suggested that the ST3 gene could be expressed in stromal cells rather than in the cancer cells themselves [9]. This hypothesis was confirmed by in situ hybridizations performed on sections of invasive breast carcinomas. In all cases ST3 RNA was only found in stromal cells (fibroblasts/myofibroblasts) surrounding the malignant epithelial cells, with the highest levels of expression in the stroma closest to cancer cell islands. Cancer cells themselves were not labeled and no significant expression could be detected in the stromal cells surrounding normal ducts and ductules.

A similar expression pattern was found by immunohistochemistry, performed with either polyclonal or monoclonal antibodies directed against ST3 (Fig. 2). ST3 protein was immunodetected in the cytoplasm of elongated fibroblastic-like cells present in the vicinity of neoplastic cells.

ST3 expression in breast pathology

Invasive carcinomas

All the invasive carcinomas examined so far by Northern blot analysis have been found to express ST3 RNA. By in situ hybridization, ST3 RNA was detected in 95% of 104 tumors [13]. The few negative cases corresponded mostly to lobular

92

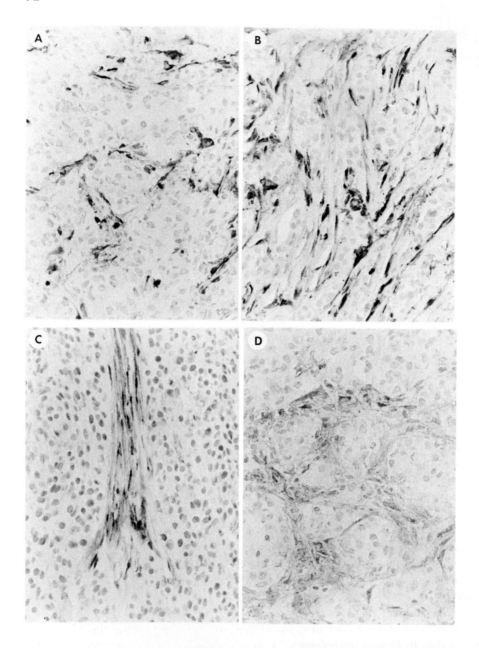

Fig. 2. Indirect immunoperoxidase staining of ST3 protein in tissue sections of invasive breast carcinomas. A) Monoclonal antibody 5ST-4A9, directed to ST3 hemopexin-like domain; ascitic fluid 1/500; frozen section. B) Monoclonal antibody 5ST-4A9; ascitic fluid 1/4000; paraffin-embedded section pretreated with microwave heating. C) Monoclonal antibody 6ST-1E11, directed to ST3 prodomain;hybridoma supernatant 1/5; frozen section. D) Polyclonal antibody 349, directed to the 25 C-terminal AA of ST3; rabbit serum 1/800; frozen section. (Original magnification: × 400).

Table 2. Comparative levels of stromelysin-3 inflammatory and noninflammatory invasive breast carcinomas[a]

ST3 positivity	% noninflammatory carcinomas (103 cases)	% inflammatory carcinomas (51 cases)
0 — weak	43/103 (42%)	11/51 (22%)
Moderate	30/103 (29%)	16/51 (31%)
High	30/103 (29%)	24/51 (47%)

[a]Analysis was performed by semiquantitative immunohistochemistry on a single section of each carcinoma.

carcinomas, which are known to have generally a less abundant stromal component than ductal carcinomas. By immunohistochemistry on frozen tissue sections with monoclonal antibody 5ST-4A9 [13] (Fig. 2), ST3 was detected in 74% of 103 infiltrative carcinomas [13], suggesting that protein detection may be slightly less sensitive than RNA detection in the evaluation of ST3 gene expression. Interestingly, an immunohistochemical study with semi-quantitative immunoscoring showed that high levels of ST3 were more frequent in inflammatory than noninflammatory carcinomas (Table 2). This observation suggests that high levels of ST3 gene expression could correlate with tumor aggressiveness.

ST3 gene expression has also been observed in breast cancer metastases. With the exception of brain metastases, which are not associated with ST3 expression, the site of metastatic tumors does not appear to be critical for ST3 gene expression. Indeed, ST3 transcripts were detected in fibroblastic cells surrounding metastatic cancer cells in lymph nodes, bone, skin, pleura and liver [13].

In situ carcinomas

ST3 transcripts and protein were also detected in fibroblastic cells of some in situ carcinomas. As in invasive carcinomas, the ST3 gene was specifically expressed in fibroblastic cells immediately surrounding the cancer cells [13]. In our series, 61% of ductal in situ carcinomas of the comedo type, but only 31% of the noncomedo types and 8% of lobular in situ carcinomas (which more rarely and more slowly evolve toward invasive tumors), expressed the ST3 gene [13] (Table 3). Moreover, both the number of ducts surrounded by fibroblastic cells expressing the ST3 gene and the levels of ST3 transcript were higher in ductal in situ carcinomas of the comedo type than in in situ carcinomas of other types. The in situ carcinomas expressing the ST3 gene were often associated with obvious remodeling of the stroma and/or questionable integrity of the basement membrane. The correlation between the frequency of ST3 gene expression in each in situ carcinoma subgroup and the known risk of these carcinomas to become invasive, suggests that ST3 may contribute to breast cancer progression from the earliest stages of tumor invasion.

Table 3. Detection of stromelysin-3 RNA in in situ breast carcinomas[a]

Type	Number of positive cases (%)	Evolution in invasion
Lobular	2/25 (8%)	+
Ductal noncomedo	7/22 (31%)	++
Ductal comedo	8/13 (61%)	+++

[a]Analysis was performed by in situ hybridization on a single section of each carcinoma.

Benign breast lesions

Most benign breast lesions do not show any ST3 gene expression. However, weak and focal expression can be observed in some sclerosing adenosis, a condition which also involves remodeling of the extracellular matrix.

ST3 expression in other human carcinomas

Although initially detected in breast carcinomas, ST3 gene expression can be observed in most other types of human carcinomas, including adenocarcinomas of colon, stomach, pancreas, prostate, endometrium and lung, and squamous cell carcinomas of cervix, skin, head and neck, and lung. In all these carcinomas, as for breast carcinomas, ST3 gene expression was always observed in fibroblastic cells of tumor stroma, with the highest levels of expression in the vicinity of cancer cells. Furthermore, in head and neck carcinomas and in skin basal cell carcinomas, high levels of ST3 expression were found to correlate with increased local invasiveness of tumors [14,15]. Most of ST3 negative cases belong to tumor types with little stroma, such as oat cell and bronchioloalveolar carcinomas of the lung, renal adenocarcinomas, and endocrine tumors.

ST3 expression in physiological conditions

Beside its expression in human carcinoma, the ST3 gene is expressed in a number of normal situations associated with tissue remodeling: during embryonic development, notably in the interdigital region of the limb buds, which corresponds to a mesenchymal tissue area associated with programmed cell death during digit individualization [9]; in syncytiotrophoblastic cells of placenta, representing the only example of ST3 epithelial expression; in stromal cells of endometrium, where ST3 has a diffuse expression pattern during the proliferative phase of the menstrual cycle, but shows limited expression under the surface epithelium and around glands during the secretory phase; in fibroblasts during skin wound healing, where ST3 expression may be under the control of factors produced by inflammatory cells [15] and during mammary involution after lactation in mice. The ST3 gene is not expressed in the mammary gland during gestation and lactation, but its expression starts 4 days after weaning, and is observed in fibroblasts immediately surrounding collapsed epithelial structures [16].

Comparative expression of ST3 with other matrix metalloproteinases and urokinase in human carcinomas

Its expression in fibroblastic cells immediately surrounding cancer cells distinguishes the ST3 gene from all other MMP genes, and in particular from the 72-kDa type IV collagenase gene which can be expressed in fibroblastic cells at a distance from the cancer cells, and which is also frequently expressed in benign tumors [13] (Table 4).

However, the expression patterns of the ST3 and urokinase gene are very similar in in situ and invasive breast carcinomas, both genes being expressed in the same population of fibroblastic cells adjacent to cancer cell islands [13]. This similarity of expression suggests that both genes may cooperate during breast cancer progression, and that their expression could be triggered by the production of a common factor emanating from the cancer cells.

Discussion

Until recently, it was generally believed that most proteolytic activities possibly involved in tumor progression were produced by cancer cells. It is now clear that several proteinases, which may play a role in tumor progression, are in fact mainly or exclusively produced by stromal cells of carcinomas, and in particular by fibroblastic cells [13,17,18] (Table 4). Among these proteinases, ST3 has unique properties. Although its substrate specificity is unknown, ST3 is expressed in fibroblastic cells of most types of human carcinomas in ways suggesting that it could be involved in tumor progression from the earliest stages of tumor invasion.

The similar expression patterns of ST3 and urokinase in breast carcinomas suggests that these two genes may co-operate during tumor progression, possibly through a proteinase cascade [19]. High urokinase levels have been shown to be potent indicators of early relapse in breast cancer patients [20–22], which suggests that the presence of ST3 could also be used to define subpopulations of tumors with aggressive potential.

Once its function has been elucidated, ST3 might also be used as a target for anticancer drugs. Indeed, as proposed by Ingber [23], targeting of noncancerous stromal cells rather than genetically unstable cancer cells may help to define new therapeutic agents with decreased likelihood of developing drug resistance.

Table 4. Comparative expression of matrix metalloproteinase and urokinase genes in breast tumors and in squamous cell carcinomas[a]

Gene	Breast fibroadenomas	Breast adenocarcinomas	Squamous cell carcinomas[b]
Interstitial type I collagenase	-	F	F+C
Gelatinase A	F	F	F
Gelatinase B	-	I+c	I+c
Matrilysin (Pump-1)	B	C	C
Stromelysin-1	-	-	-
Stromelysin-2	-	c	C
Stromelysin-3	-[c]	F	F
Urokinase	-	F+c	C

[a]Analysis by RNA in situ hybridization; [b]From head and neck, and skin; [c]ST3 RNA was detected in fibroblasts of one fibroadenoma out of 21 examined [13].
I = inflammatory cells; F = fibroblastic cells; C = cancer cells; c = few cancer cells, B = benign epithelial cells; - = no expression.

Acknowledgements

We would like to thank B. Peltre and B. Heyd for technical assistance, Y. Lutz for preparing the monoclonal antibodies, and J. Byrne for critically reading the manuscript. This work was supported by funds from the Centre National de la Recherche Scientifique, the Institut National de la Santé et de la Recherche Médicale, the Mutuelle Générale de l'Education Nationale, le Ministère de la Recherche et de l'Espace (contract No. 92H.0917), the Centre Hospitalier Universitaire Régional, the Association pour la Recherche sur le Cancer, la Ligue Nationale Française contre le Cancer, la Fondation pour la Recherche Médicale Française and la Fondation Jeantet.

References

1. Liotta A, Stetler-Stevenson WG. Cancer Res 1991;51(suppl):5054s–5059s.
2. Duffy MJ. Blood Coag Fibrinol 1990;1:681–687.
3. Docherty AJP, O'Connel J, Crabbe T, Angal S, Murphy G. Tibtech 1992;10:200–207.
4. Kwaan HC. Cancer Metastasis Rev 1992;11:291–311.
5. Matrisian LM. BioEssays 1992;14:455–463.
6. Liotta LA, Nageswara RC, Wewer UM. Ann Rev Biochem 1986;55:1037–1057.
7. Blood CH, Zetter BR. Biochim Biophys Acta 1990;1032:89–118.
8. Fidler IJ. Cancer Res 1990;50:6130–6138.
9. Basset P, Bellocq JP, Wolf C, Stoll I, Hutin P, Limacher JM, Podhajcer OL, Chenard MP, Rio MC, Chambon P. Nature 1990;348:699–704.
10. Murphy GFP, Murphy G, Reynolds JJ. FEBS Lett 1991;289:4–7.
11. Murphy G, Segain JP, O'Shea M, Cockett M, Ioannou C, Lefebvre O, Chambon P, Basset P. J Biol Chem 1993;268:15435–15441.
12. Murphy G, Allan JA, Willenbrock F, Cockett MI, O'Connell JP, Docherty AJP. J Biol Chem 1992; 267:9612–9618.

13. Wolf C, Rouyer N, Lutz Y, Adida C, Loriot M, Bellocq JP, Chambon P, Basset P. Proc Natl Acad Sci USA 1993;90:1843–1847.
14. Muller D, Wolf C, Abecassis J, Millon R, Engelmann A, Bronner G, Rouyer N, Rio MC, Eber M, Methlin G, Chambon P, Basset P. Cancer Res 1993;53:165–169.
15. Wolf C, Chenard MP, Durand de Grossouvre P, Bellocq JP, Chambon P, Basset P. J Invest Dermatol 1992;99:870–872.
16. Lefebvre O, Wolf C, Limacher J, Hutin P, Wendling C, LeMeur M, Basset P, Rio MC. J Cell Biol 1992;119:997–1002.
17. Grondahl–Hansen J, Ralfkiaer E, Kirkeby LT, Kristensen P, Lund LR, Dano K. Am J Pathol 1991; 138:111–117.
18. Pyke C, Ralfkiaer E, Tryggvason K, Dano K. Am J Pathol 1993;142:359–365.
19. Mignatti P, Robbins E, Rifkin DB. Cell 1986;47:487–498.
20. Duffy MJ, Reilly D, O'Sullivan C, O'Higgins N, Fennelly JJ, Andreasen P. Cancer Res 1990;50: 6827–6829.
21. Jänicke F, Schmitt M, Hafter R, Hollrieder A, Babic R, Ulm K, Gössner W, Graeff H. Fibrinolysis 1990;4:1–10.
22. Grondahl-Hansen J, Christensen IJ, Rosenquist C, Brünner N, Mouridsen HT, Dano K, Blichert-Toft M. Cancer Res 1993;53:2513–2521.
23. Ingber DE. Sem Cancer Biol 1992;3:57–63.

Prospects in diagnosis and treatment of breast cancer
M. Schmitt et al., editors

The role of cell adhesion molecules in metastasis formation by solid tumors

Judith P. Johnson
Institute for Immunology, Goethestrasse 31, 80336 Munich, Germany

Abstract. Cell adhesion molecules play important roles at several stages in tumor progression. The downregulation of cell adhesion molecules expressed by normal epithelial cells is one of the earliest occurrences observed in carcinomas. However, expression of new cell adhesion molecules also characterizes tumor progression. All those which have been identified to date on carcinomas and neuroectodermally derived tumors are molecules normally expressed on leukocytes or endothelia and which function to direct leukocyte trafficking. Expression of these molecules is often associated with poor survival, suggesting that tumors use these pathways in their dissemination.

Introduction

Metastasis formation is a complex process in which the tumor cells must separate from the primary tumor mass, invade the surrounding tissues, travel through the vascular system, re-enter the tissues and establish foci of growth in a new surrounding. The development of metastatic capacity by the tumor cells appears to be a stepwise process and is associated with the acquisition of a number of new characteristics, including the autocrine production of growth factors and the production of substrate adhesion molecules and proteolytic enzymes which enable the tumor cells to move through the extracellular matrix [1,2]. In recent years evidence has been accumulating that molecules mediating adhesion between cells play important roles at several stages in this process.

Cell adhesion molecules (CAMs) are cell surface molecules which bind cells together. There are two general types of CAMs, heterophilic and homophilic. A homophilic CAM interacts with itself, i.e., a CAM on one cell binds to the same molecule on a neighboring cell. Homophilic CAMs generally mediate cell adhesion between cells of the same type (known as homotypic adhesion) and are of critical importance in the establishment and maintenance of normal tissue architecture [3]. A heterophilic CAM interacts with a different CAM on the neighboring cell. Heterophilic CAMs frequently mediate adhesion between two distinct cell types. Perhaps the best known examples of heterotypic cell adhesion are in the immune system [4]. Although a T lymphocyte recognizes its target through the interaction of its T cell receptor with an antigen complex on the target cell, the result of this recognition — be it stimulation of the T cell or killing of the target cell — generally only occurs when interactions between CAMs on the two cells also take place. In fact, CAM interactions are frequently accompanied by the transmission of signals regulating differentiation and proliferation to one or both partners. Even more importantly, by maintaining cells in close contact, CAMs allow the exchange of information mediated by other cellular components to occur. In this context, CAMs play an important role in virtually all aspects of intercellular communication in multicellular organisms.

The malignant cell demonstrates a severely altered relationship with its environment. Connections to neighboring cells are disturbed and tumor cells migrate away from their normal location to take up residence in foreign environments. It is therefore not surprising that disturbances in CAM function should be associated with malignant transformation and tumor progression.

Tumors frequently demonstrate a loss of normally expressed homophilic CAMs

One of the earliest events in the formation of metastasis is the separation of single tumor cells or clusters of tumor cells from the primary tumor. There is increasing evidence that this occurs, at least in part, through an alteration of normal cell — cell adhesion and the consequent loss of contact-mediated regulation.

One of the molecules which has been implicated most frequently in this process is E-cadherin, a member of a large family of calcium-dependent cell adhesion molecules which mediate homophilic adhesion [5]. E-cadherin is expressed on the surface of essentially all normal epithelia. However, a loss of E-cadherin expression has been observed in situ in many different human carcinomas. Mayer et al. [6] observed a downregulation of E-cadherin, reflected both in terms of the level of expression and in a heterogeneity of expression, in 92% of 60 primary gastric carcinomas examined. Similar observations have also been made for colorectal carcinomas [7], head and neck tumors [8], thyroid carcinomas [9], prostate [10], bladder [11] and breast carcinomas [12,13]. The presence or absence of E-cadherin is thought to contribute to the "invasive" character of the tumor cell since its downregulation in cell lines leads to an increase in their ability to invade the extracellular matrix, a characteristic which can be reversed by the transfection of E-cadherin cDNA [14,15]. In all human tumors examined, the loss or downregulation of E-cadherin expression was found to correlate with the loss of differentiated characteristics of the tumors. Given the well documented prognostic significance of the loss of differentiated phenotype, it is therefore not surprising that the loss of E-cadherin expression has also been shown to correlate with reduced disease-free interval and poor survival [6,11]. E-cadherin is not the only CAM which is lost or inactivated during tumor development; similar observations have been made for DCC in colorectal carcinomas [16] and for NCAM in Wilms tumor [17].

Table 1. Leukocyte CAMs expressed by solid tumors

CAM	Tumor type	Endothelial ligand	Reference
VLA-4	melanoma	VCAM-1	[27]
CD44var	colon, gastric, breast, ovary	?	[23,25,28]
sLewis^x	colon, gastric, ovary, lung, pancreas, liver	E-selectin	[22]
HNK-1(CD57)	melanoma, neuroblastoma, SCLC	P-selectin	[29]
Sulfatid	melanoma, breast	P-selectin	[30]

SCLC = small cell lung carcinoma.

Tumors frequently demonstrate de novo expression of heterophilic CAMs

During the process of tumor progression tumor cells do not only lose their connections to their neighbors, but also establish new interactions as they travel through the vascular system and take up residence in new environments. Just as the loss of normal cell contacts can be reflected in changes in CAM expression, many of these new interactions are also likely to be mediated at least in part by CAMs. And while the loss of normal interactions is associated with the loss of homotypic cell adhesion, the acquisition of new characteristics might be expected to be associated with the expression of new CAMs which mediate heterotypic interaction.

In recent years, evidence has been accumulating that tumor progression is in fact accompanied by the de novo expression of a variety of CAMs. As these molecules are identified, it is becoming increasingly clear that most are heterophilic CAMs which are normally expressed in the hematopoietic system where they mediate the trafficking of leukocytes between the vascular system and the tissues.

The migration of leukocytes from the vascular system into the tissues is orchestrated by changes in the expression of a variety of heterophilic endothelial-leukocyte CAMs [18]. Under conditions of normal blood flow, leukocyte "catching" in areas of inflammation is mediated by 2 cytokine-inducible endothelial CAMs of the selectin family, P-selectin and E-selectin [19]. These lectin-like CAMs interact with carbohydrate ligands on leukocytes and although this interaction is transient, it serves to slow the leukocytes and induces their "rolling" along the endothelium. Tight adhesive bonds are then established between the leukocytes and the endothelial molecules ICAM-1 and VCAM-1, and the cells are transported across the endothelium and into the tissues [20]. While the expression of the ligands of these endothelial CAMs is normally limited to subpopulations of leukocytes, several of these leukocyte CAMs have now been found on carcinomas and neuroectodermal derived tumors (Table 1). These include the integrin VLA-4, various carbohydrate epitopes which serve as ligands for the endothelial selectins, and lymphocyte-specific forms (variants or isoforms) of CD44, a highly glycosylated cell surface molecule which has also been shown to play a role in leukocyte-endothelial interactions [21]. In vitro studies have confirmed that the presence of the "leukocyte CAM" confers endothelial binding capability to the tumor cell [22–24], strongly suggesting that the tumor cells are able to use these molecules in vivo to cross the endothelium and enter the tissues. Consistent with this is the observation that expression of these molecules by the tumors can sometimes be shown to be a prognostic marker correlating with poor survival [25,26].

A second group of heterophilic CAMs which are being identified on solid tumors are also involved in leukocyte-endothelial interactions, but these are molecules which are normally expressed by the endothelial partner (Table 2). In cutaneous melanoma the development of metastatic potential is strongly correlated with the vertical depth of the primary tumor. The isolation of monoclonal antibodies reactive with advanced tumors, but not with early tumors and benign lesions, is an approach which can lead to the identification of molecules which are up regulated or newly expressed as tumors develop metastatic potential. In our laboratory, two such molecules, P3.58 and MUC18 were identified using this approach (Fig. 1). cDNA cloning revealed that both are cell surface glycoproteins belonging to the immunoglobulin super family and both appear to be CAMs. P3.58 was found to be identical to ICAM-1, a cell adhesion molecule mediating adhesion with leukocytes [34]. MUC18 is a unique gene but on the basis of structural similarity to a number of well defined CAMs, it is also likely to be a CAM [35]. Both MUC18 and ICAM-1 have been found to be expressed on endothelia in situ, particularly in areas of inflammation and in tumors. Analysis of protein and mRNA from endothelial cells in vitro confirmed that both ICAM-1 and MUC18 are expressed by these cells (Sers et al. submitted). The similar expression patterns of MUC18 and ICAM-1 on endothelia in vitro and in vivo, suggest that MUC18 also interacts with a leukocyte ligand and that it may be important in the circulation and homing of these cells into the tissues.

The expression of endothelial CAMs on solid tumors raises the question of whether the tumors may also be using these molecules to enhance their hematogenous spread. A prospective study has shown that the expression of ICAM-1 by stage I melanomas does in fact have a prognostic significance [36]. Patients with ICAM-1 positive tumors had a significantly shorter disease-free interval and survival time than

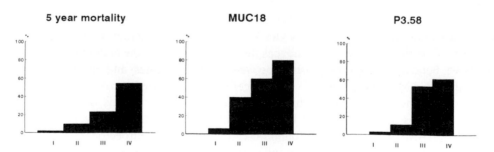

Fig. 1. Correlation between the expression of MUC18 and P3.58 on primary cutaneous melanomas and the 5 year mortality rate. Percentage reactive lesions are shown on the y-axis and the vertical depth of the tumor on the x-axis (I, <0.75 mm; II, 0.76-1.5 mm; III, 1.51-3.0 mm; IV >3.0 mm.

Table 2. Endothelial CAMs expressed by solid tumors

CAM	Tumor type	Leukocte ligand	Reference
ICAM-1	melanoma, breast, kidney	CD11a, CD11b, CD43	[31]
VCAM-1	melanoma	VLA-4	[32]
MUC18	melanoma, neuroblastoma, SCLC	?	[33]

patients with ICAM-1 negative tumors. While such studies are not yet available regarding MUC18, a positive correlation has been observed between the expression of this molecule by human tumor cells and the ability of these cells to generate metastases in nude mice [37]. Thus there is indirect evidence that both ICAM-1 and MUC18 expression by tumor cells can enhance the probability of metastasis generation.

As the ligands of endothelial CAMs are normally expressed by leukocytes, tumor cells expressing these molecules would be expected to aggregate with leukocytes, and for ICAM-1 on melanoma cells this has been directly demonstrated [38]. A number of in vivo studies in animal models have shown that tumor cell-leukocyte aggregates are more efficient in the generation of both spontaneous and experimental metastatic lesions than are tumor cells alone [39,40]. The presence of leukocytes may provide a variety of growth factors for the tumors as well as proteolytic enzymes which aid the tumor cells in migrating through the extracellular matrix. In addition, such tumor cell-leukocyte aggregates in the vascular systems not only serve to protect the tumor cells from sheer damage but can provide a vehicle (that is the leukocyte-endothelial interactions) by which the tumor cells can migrate into the tissues.

If carcinomas and neuroectodermally derived tumors actually do use the normal leukocyte trafficking molecules in their dissemination, the various therapeutic approaches which are being developed to interfere with leukocyte entry into tissues in situations such as graft rejection [41], asthma [42] and myocardial infarction [43] will hold an even more exciting possibility; they may also be able to prevent or reduce metastasis formation if used in the early stages of tumor growth.

Acknowledgements

This work was supported by a grant from the Deutsche Krebshilfe, Mildred Scheel Stiftung, Bonn, Germany.

References

1. Fidler IJ. Cancer Res 1990;50:6130–6138.
2. Hart IR, Saini A. Lancet 1992;339:1453–1457.
3. Edelman GM, Crossin KL. Ann Rev Biochem 1991;60:155–190.

4. Springer TA. Nature 1990;346:425–434.
5. Takeichi M. Ann Rev Biochem 1990;59:237–252.
6. Mayer B, Johnson JP, Leitl F, Jauch KW, Heiss MM, Schildberg FW, Birchmeier W, Funke I. Cancer Res 1993;53:1690–1695.
7. Dorudi S, Sheffield JP, Poulsom R, Northover JMA, Hart IR. Am J Pathol 1993;142:981–986.
8. Schipper JH, Frixen UH, Behrens J, Unger A, Janke K, Birchmeier W. Cancer Res 1991;51:6328–63–37.
9. Brabant G, Hoang-Vu C, Cetin Y, Dralle H, Scheumann G, Mölne J, Hansson G, Jansson S, Ericson LE, Nilsson M. Cancer Res 1993;53:4987–4993.
10. Umbas R, Schalken JA, Aalders TW, Carter BS, Karthaus HFM, Schaafsma HE, Debruyne FMJ, Issacs WB. Cancer Res 1992;52:5104–5109.
11. Bringuier PP, Umbas R, Schaafsma HE, Karthaus HFM, Debruyne FMJ, Schalken JA. Cancer Res 1993;53:3241–3245.
12. Gamallo C, Palacios J, Suarez A, Pizarro A, Navarro P, Quintanilla M, Cano A. Am J Pathol 1993; 142:987–993.
13. Oka H, Shiozaki S, Kobayashi K, Inoue M, Tahara H, Kobayashi T, Takatsuka Y, Matsuyoshi N, Hirano M, Takeichi M, Mori T. Cancer Res 1993;53:1696–1701.
14. Vleminckx K, Vakaet Jr L, Mareel M, Fiers W, Van Roy F. Cell 1991;66:107–119.
15. Frixen UH, Behrens J, Sachs M, Eberle G, Voss B, Warda A, Löchner D, Birchmeier W. J Cell Biol 1991;113:2435–2447.
16. Fearon ER, Cho KR, Nigro JM, Kern SE, Simons JW, Ruppert JM, Hamilton SR, Preisinger AC, Thomas G, Kinzler KW, Vogelstein B. Science 1990;247:49–56.
17. Roth J, Zuber C, Wagner P, Taatjes DJ, Weisgerber C, Heitz PU, Gordis C, Bitter-Suermann D. Proc Natl Acad Sci USA 1988;85:2999–3003.
18. Osborn L. Cell 1990;62:3–6.
19. Bevilacqua MP, Nelson RM. J Clin Invest 1993;91:379–387.
20. Butcher E. Cell 1991;67:1033–1036.
21. Arch R, Wirth K, Hofmann M, Ponta H, Matzku S, Herrlich P, Zöller M. Science 1992;257:682–685.
22. Takada A, Ohmori K, Yoneda T, Tsuyuoka K, Hasegawa A, Kiso M, Kannagi R. Cancer Res 1993; 53:354–361.
23. Cannistra SA, Kansas GS, Niloff J, DeFranzo B, Kim Y, Ottenmeier C. Cancer Res 1993;53: 3830–3838.
24. Rice GE, Bevilacqua MP. Science 1989;1303:1306.
25. Mayer B, Jauch KW, Günthert U, Figdor CG, Schildberg FW, Funke I, Johnson JP. Lancet 1993;342:1019–1022.
26. Nakamori S, Kameyama M, Imaoka S, Furukawa H, Ishikawa O, Sasaki Y, Kabuto T, Iwanaga T, Matsushita Y, Irimura T. Cancer Res 1993;53:3632–3637.
27. Martin-Padura I, Mortarini R, Lauri D, Bernasconi S, Sanchez-Madrid F, Parmiani G, Mantovani A, Anchini A, Dejana E. Cancer Res 1991;51:2239–2241.
28. Heider KH, Hoffmann M, Horst E, van den Berg F, Ponta H, Herrlich P, Pals ST. J Cell Biol 1993;120:227–233.
29. Needham LK, Schnaar RL. Proc Natl Acad Sci USA 1993;90:1359–1363.
30. Aruffo A, Kolanus W, Walz G, Fredman P, Seed B. Cell 1991;67:35–44.
31. Johnson JP. Chem Immunol 1991;50:143–161.
32. Jonjiic N, Martin-Padura I, Pollicino T, Bernasconi S, Jilek P, Bigotti A, Morarini R, Anichini A, Parmiani G, Colotta F, Dejana E, Mantovani A, Natali PG. Am J Path 1992;141:1323–1330.
33. Sers C, Kirsch K, Rothbächer U, Riethmüller G, Johnson JP. Proc Natl Acad Sci USA 1993;90: 8514–8518.
34. Johnson JP, Stade BG, Holzmann B, Schwäble W, Riethmüller G. Proc Natl Acad Sci USA 1989; 86:641–644.
35. Lehmann JM, Riethmüller G, Johnson JP. Proc Natl Acad Sci USA 1989;86:9891–9895.

36. Natali P, Nicotra MR, Cavaliere R, Bigotti A, Romano G, Temponi M, Ferrone S. Cancer Res 1990; 50:1271—1278.
37. Luca M, Hunt B, Bucana CD, Johnson JP, Fidler IJ, Bar-Eli M. Mel Res 1993;3:35—41.
38. Anichini A, Mortarini R, Alberti S, Mantorani A, Parmiani G. Int J Cancer 1993;53:994—1001.
39. Starkey JR, Liggitt HD, Jones W, Hosick HL. Int J Cancer 1984;34:535—543.
40. Blood CH, Zetter BR. Biochem Biophys Acta. 1990;1032:89—118.
41. Isobe M, Yagita H, Okumura K, Ihara A. Science. 1992;255:1125—1127.
42. Wegner CD, Gundel RH, Reilly P, Haynes N, Letts LG, Rothlein R. Science 1990;247:456—459.
43. Simpson PJ, Todd RF, Fantone JC, Mickelson JK, Griffin JD, Lucchesi BF. J Clin Invest 1988;81: 624—629.

The invasion suppressor function of E-cadherin in mammary epithelioid cells

Marc E. Bracke[1]*, Stefan J. Vermeulen[1], Erik A. Bruyneel[1], Krist'l M. Vennekens[1], Georges K. De Bruyne[1], Frans M. van Roy[2] and Marc M. Mareel[1]

[1]*Laboratory of Experimental Cancerology, Department of Radiotherapy, Nuclear Medicine and Experimental Cancerology, University Hospital, De Pintelaan 185, B-9000 Gent; and [2]Laboratory of Molecular Biology, K.L. Ledeganckstraat 35, B-9000 Gent, Belgium*

Abstract. MCF-7/6 human mammary carcinoma cells are invasive in vitro and in vivo. These cells show poor cell-cell adhesion due to dysfunction of E-cadherin at the cell surface. Three molecules have been found to restore the defect of E-cadherin function: the hormone insulin-like growth factor I, the citrus flavonoid tangeretin and the morphogen retinoic acid. These molecules stimulate MCF-7/6 cell-cell adhesion and inhibit invasion in embryonic chick heart fragments in vitro. These findings support the opinion that E-cadherin is an invasion suppressor molecule.

Cell-cell adhesion has long since been recognized as a key mechanism in the maintenance of positional stability of normal epithelial cells [1]. The discovery and characterization of cell-cell adhesion molecules during the last decade has expanded this research area substantially [2,3]. E-cadherin is an example of such a molecule, which is present at the surface of probably all types of normal epithelial cells, and which is known to establish cell-cell adhesion via homophilic (between like molecules) and mostly homotypic (between like cells) interactions [4]. This 120-kDa molecule consists of three domains. The extracellular domain contains cell binding regions, which depend on calcium for functional integrity, and a number of glycosylation sites. A small membrane-spanning domain connects the extracellular domain with the cytoplasmic one, which is in its turn noncovalently coupled to the (actin) cytoskeleton via the catenins. Both E-cadherin and catenins possess phosphorylation sites [5,6].

Invasion of malignant epithelioid cell populations (carcinomas) has in many instances been related to the loss of E-cadherin-mediated cell-cell adhesion. Many immunohistochemical studies have demonstrated an inverse correlation between invasion and E-cadherin expression in tumor samples, including mammary carcinomas [7–9]. Experiments in vitro have revealed the functional implications of E-cadherin expression by using constitutively invasive E-cadherin-negative cells, which could be

---

Address for correspondence: Laboratory of Experimental Cancerology, Department of Radiotherapy, Nuclear Medicine and Experimental Cancerology, University Hospital, De Pintelaan 185, B-9000 Gent, Belgium.

rendered noninvasive via introduction of the E-cadherin gene and efficient expression of the molecule at the plasma membrane [10–12]. Alternatively, in constitutively noninvasive E-cadherin-positive epithelial cells, invasion was observed after neutralization of the E-cadherin function with antibodies [13] or after reducing its expression with antisense RNA [10]. Together these observations have led to considering E-cadherin as an invasion suppressor molecule.

The lack of homogeneous E-cadherin expression in particular experimental carcinomas has been attributed to host-derived downregulating factors from the microenvironment. The expression of E-cadherin can transiently be downregulated by passage in laboratory animals and resumes, often rapidly, in explants of these tumors ex vivo [14]. The cellular origin and the nature of such presumptive factors, however, are unknown so far. Another mechanism, based on the expression of functionally inactive E-cadherin, has been proposed by us to explain defective cell-cell adhesion and invasion. For this we examined a number of variants of the human MCF-7 cell line, which was originally derived from the pleural effusion of a metastasizing mammary adenocarcinoma [15]. These variants were obtained from different laboratories, and their MCF-7 origin was confirmed by specific immunoreactive markers [16]. All variants were confronted in vitro with embryonic chick heart fragments, cultured in suspension for 8 days, and their interaction with the chick heart fragments were evaluated by histology [17]. Some variants (e.g., MCF-7/6) invaded into the heart tissue, while others (e.g., MCF-7/AZ) failed to do so [18]. The presence of E-cadherin and the catenins at the plasma membrane of both MCF-7/6 and MCF-7/AZ cells was demonstrated via multiple techniques (immunocytochemistry,

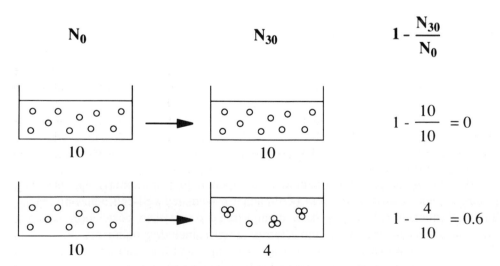

$$N_0 \qquad N_{30} \qquad 1 - \frac{N_{30}}{N_0}$$

$$1 - \frac{10}{10} = 0$$

$$1 - \frac{4}{10} = 0.6$$

Fig. 1. Schematic representation of the fast aggregation assay. Cells are brought in suspension under E-cadherin-saving conditions, and the number of particles is counted initially (N_0) and after 30 min (N_{30}). The expression $1-N_{30}/N_0$ is a measure for the tendency of the cells to aggregate. $1-N_{30}/N_0$ approaches 0 with nonaggregating cells (top), and increases up to a theoretical maximum of 1 with aggregating cells (bottom).

immunohistochemistry, flow cytometry and Western blotting) [19]. From these data we concluded that E-cadherin was unable to exert its invasion suppressor function in MCF-7/6 cells.

Comparison between invasive MCF-7/6 and noninvasive MCF-7/AZ cells revealed differences in cell-cell adhesion in vitro. Our fast aggregation assay is inspired by the work of Kadmon et al. [20]. Cells are detached from their plastic tissue culture substrate under conditions that should leave E-cadherin functionally intact. By Coulter counter measurements, the initial number of particles in suspension is compared with the number after 30 min of aggregation [19] (Fig. 1). In this assay MCF-7/6 cells showed poor fast aggregation in contrast with their noninvasive counterparts. MCF-7/AZ aggregation was specifically abolished in the presence of monoclonal antibodies directed against a functional domain of E-cadherin (HECD-1 [21], MB2 [19]), but not by other monoclonals binding to other epitopes of proteins at the plasma membrane (e.g., 5D10 [22]). These results indicate that E-cadherin at the surface of MCF-7/6 cells is not functionally active. One of the goals of our current research is to find agents that are able to correct E-cadherin function in the invasive MCF-7/6 variant as far as cell-cell adhesion and invasion suppression are concerned.

Insulin-like growth factor I can correct the defective function of E-cadherin in MCF-7/6 human mammary carcinoma cells

Insulin-like growth factor I (IGF-I) is a 7-kD polypeptide hormone with multiple functions, as reflected by the many names that were attributed to this molecule in the past: sulphation factor, nonsuppressible insulin-like activity-soluble (NSILA-S), multiplication stimulating activity (MSA) and somatomedin C [23]. Secreted by hepatocytes under the regulation of growth hormone, IGF-I is released into the blood, and can act as an endocrine factor. However, more recently, short-range paracrine [24] and autocrine [25] effects have been elucidated. We have shown that IGF-I can correct the defective cell-cell adhesion function of E-cadherin in the invasive MCF-7/6 cell variant at near-physiological serum concentrations [19]. Fast aggregation is stimulated by IGF-I within minutes, and the effect is independent from de novo protein synthesis. This increased aggregation is mediated by E-cadherin, since it can be blocked by the monoclonal anti-E-cadherin antibodies HECD-1 and MB2.

IGF-I interacts with the IGF-I receptor on the MCF-7 cell surface [26]. This receptor, a heterotetramer of 2 α and 2 β subunits, is present on MCF-7/6 cells, as evidenced by us by Scatchard analysis, and by flow cytometry and immunoprecipitation with the anti-IGF-I-receptor monoclonal antibody αIR3. Furthermore, the latter monoclonal antibody inhibited completely the effect of IGF-I on MCF-7/6 aggregation. MCF-7 cells also express insulin receptors with homology to the IGF-I receptor [27], and insulin is indeed able to mimic the IGF-I effect on cell aggregation, albeit at supraphysiological concentrations. The insulin effect was not blocked by the monoclonal antibody αIR3, indicating that insulin does not act via the IGF-I receptor, but presumably via its own receptor. It is not excluded that our cells possess a hybrid of the IGF-I and the insulin receptor, which is less sensitive to IGF-I and insulin than the proper IGF-I or insulin receptors respectively [28]. Insulin-like growth factor II (IGF-II), having its own type of receptor, was not able to induce fast cell aggregation of MCF-7/6 cells.

Signal transduction from the triggered IGF-I receptor starts with switching on its intracellular tyrosine kinase activity, which leads to autophosphorylation and to phosphorylation of (a) cytoplasmic substrate(s) [29,30]. For the insulin receptor these phosphorylations have recently been shown to result in *ras* activation through a cascade of rapid molecular interactions [31,32]. We have shown that activation of E-cadherin-mediated aggregation of MCF-7/6 cells, can be blocked by a number of tyrosine kinase inhibitors: genistein (25 µM), me 2,5-diOH cinnamate (50 µM) and 2-OH-5-(2,5 diOH benzyl) aminobenzoic acid (10 µM). Our current research aims at revealing possible effects of IGF-I on the phosphorylation of catenins and E-cadherin. Phosphorylation of catenins has recently been shown to modulate the function of E-cadherin [5,6].

Extracellular regulation of IGF-I as a trigger for its receptor is mediated by the insulin-like growth factor binding proteins (IGFBPs). These proteins are present in blood [33] and are also secreted by MCF-7 cells in their culture medium. Six species of IGFBPs have been characterized, five of which can be secreted by MCF-7 cells [34]. Ligand blotting of MCF-7/6 conditioned media with ^{125}I-labelled IGF-I revealed at least three bands (Fig. 2). While in a few cases binding of IGF-I with an IGFBP was reported to result in an increase of IGF-I potency, such a binding usually prevents the binding of the ligand to its receptor [34].

The impact of IGFBPs on the correction of E-cadherin-mediated cell-cell adhesion function via IGF-I, can be illustrated by the potency of (des 1-3) IGF-I. The latter is a truncated variant of IGF-I, which lacks the first three N-terminal residues [35]. This variant has lost most of its binding affinity for IGFBPs, and is about 100 times more potent to activate E-cadherin-mediated fast aggregation of MCF-7/6 cells, as compared with nontruncated IGF-I (Fig. 3). A possible autocrine loop via this

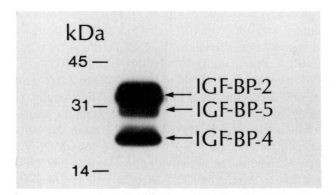

Fig. 2. Ligand blotting of IGF-binding proteins secreted by MCF-7/6 cells in the culture medium. SDS-PAGE of conditioned medium was blotted and revealed with 125I-labelled IGF-I.

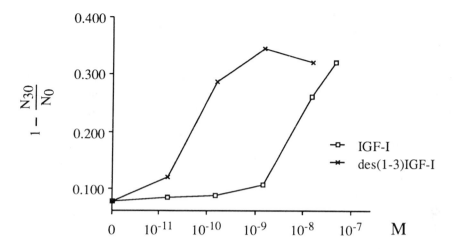

Fig. 3. Effect of (des 1-3) IGF-I and of nontruncated IGF-I on MCF-7/6 fast cell aggregation (see Fig.1 for aggregation index $1-N_{30}/N_0$). (des 1-3) IGF-I is about 100x more potent than native IGF-I to stimulate cell-cell adhesion between MCF-7/6 cells. This difference in potency is most probably due to the reduced binding of (des 1-3) IGF-I to IGF-binding proteins in the aggregation medium.

truncated IGF-I is not excluded in some types of MCF-7 cells, since (des 1—3) IGF-I has been shown to be produced by at least some MCF-7 cells [36].

When added to the medium of confronting cultures of MCF-7/6 cells with embryonic chick heart fragments, IGF-I exerted an anti-invasive effect, which was reversible upon omission of the factor [19]. Again this effect appeared to be mediated by the IGF-I receptor, because it could be blocked by a monoclonal antibody against the IGF-I receptor. Our finding that IGF-I receptor triggering prevents invasion in vitro should be compared with studies on human mammary tissue samples indicating production of IGF-I by stromal cells of nonmalignant specimens, and production of IGF-II by stromal cells of invasive carcinomas [37,38]. Recently it was also found that expression of high levels of IGF-I receptor in mammary tumors is an indicator of better prognosis [39,40].

The citrus flavonoid tangeretin can correct the defective function of E-cadherin via a mechanism that is not mediated by the IGF-I receptor

Flavonoids are interesting tools to study the mechanisms of mammary tumor invasion. Examples are the weak estrogen and strong tyrosine kinase inhibitor genistein [41,42], the pp60[src] inhibitor quercetin [43], the laminin-binding anti-invasive flavonol (+)–catechin [44] and the biological response modifier flavone acetic acid [45]. Tangeretin is a flavonoid extractable from citrus plants, which was retained in a screening program for potential anti-invasive molecules in the assay with embryonic chick heart [46]. This molecule (Fig. 4) inhibits the invasion of MO4 virus-

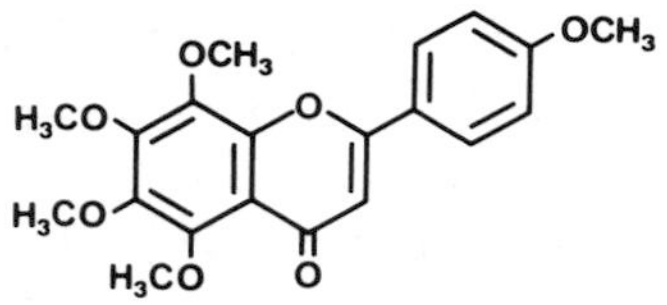

Fig. 4. Structure of tangeretin.

transformed mouse cells and of MCF-7/6 cells. As a polymethoxylated flavonoid, tangeretin is stable both in vitro and in vivo. After oral administration via drinking water, it is found in the liver of laboratory mice [47].

Tangeretin corrects the defective function of E-cadherin in MCF-7/6 cells in a way that resembles the effect of IGF-I. Tangeretin induces MCF-7 fast cell aggregation. This phenomenon is mediated by E-cadherin, since it can be blocked by the anti-E-cadherin monoclonal antibody HECD-1 and MB2, and it is not dependent on de novo protein synthesis, since cycloheximide does not inhibit the tangeretin effect. The flavonoid effect, however, differs from the IGF-I effect in that it does not require triggering of the IGF-I receptor, as evidenced by the lack of effect of αIR3, the monoclonal antibody that inhibits the function of this receptor.

Experiments with estrogen-primed female nude mice are going on to explore whether tangeretin can prevent liver metastases from MCF-7 tumors raised in the spleen [48]. Such experiments are also inspired by the effect of flavone acetic acid on the development of spleen-derived artificial liver metastases [49].

Retinoic acid induces fast aggregation of MCF-7/6 cells

All-*trans*-retinoic acid (RA) is a vitamin A derivative with opposite effects on invasion depending on the type of tumor cells under study. When added to the medium of confronting cultures with embryonic chick heart fragments, RA inhibits invasion of MCF-7/6 cells, but induces invasion in the constitutively noninvasive cell variant MCF-7/AZ [18]. RA affects multiple functions of MCF-7/6 cells: it decreases fast plasma membrane motility, it increases the enzymatic activity of tissue-type transglutaminases [50] and it increases cell-cell adhesion. So, the anti-invasive effect of RA on MCF-7/6 can be explained by inducing a less motile, more rigid and more strongly adherent cell population.

Fast aggregation of MCF-7/6 cells was increased by addition of RA to the medium. This effect was about maximal after pretreatment of the cells during 4 h, and was concentration-dependent. The anti-E-cadherin monoclonal antibodies HECD-1 and MB2 were able to block the effect of RA on cell aggregation, which indicates that RA specifically corrects the defective function of E-cadherin in these cells. The RA effect was not sensitive to cycloheximide, which shows that de novo protein synthesis is not required.

The action mechanism of RA on MCF-7/6 cells is not well understood. RA is considered to be a morphogen [51,52], that penetrates the plasma membrane, and is transported in the cytoplasm towards the nucleus via the cytoplasmic RA-binding protein (CRABP). This CRABP is present in MCF-7/6 cells [50]. Interaction with the nuclear RA receptors modulates gene transcription and other molecules can interact with this zinc-finger mediated phenomenon [53]. Cycloheximide insensitivity makes classical promotion of DNA transcription/translation of relevant genes by RA rather improbable. Other possible action mechanisms, however, are now being studied, such as the effect of RA on cytoplasmic protein tyrosine phosphatases. The observation that vanadate treatment can mimic RA effects on cell behavior [54] may indicate that RA inhibits protein tyrosine phosphatase activity. Such a mechanism could induce a higher level of tyrosine phosphorylation of the IGF-I receptor and of its intracellular substrates, and could tune up the sensitivity of the receptor for ambient IGF-I. At least one observation seems to indicate that the triggering of IGF-I receptor is implicated in the action mechanism of RA: blocking of the IGF-I receptor by the monoclonal antibody αIR3 inhibits the effect of RA on fast aggregation.

Conclusion

IGF-I, tangeretin and retinoic acid are inhibitors of MCF-7/6 invasion into embryonic chick heart in organ culture. These three molecules also increase E-cadherin-mediated cell-cell adhesion. Our results strengthen the opinion that E-cadherin is an invasion suppressor in MCF-7 cells. Signal transduction from these molecules towards E-cadherin is the topic of our ongoing research.

Acknowledgements

We thank Jean Roels van Kerckvoorde for preparing the illustrations. This work was supported by grants from the Department of Citrus of the State of Florida USA, from the Vereniging voor Kankerbestrijding and from the Onderzoeksfonds van de ASLK (36 1131 88). Frans M. van Roy is Research Director with the Belgian N.F.W.O.

References

1. Lewis WH. Anat Rec 1922;23:387–392.
2. Edelman GM, Crossin KL. Ann Rev Biochem 1991;60:155–190.
3. Takeichi M. Science 1991;251:1451–1455.
4. Kemler R. Sem Cell Biol 1992;3:149–155.
5. Behrens J, Vakaet L, Friis R, Winterhagen E, Van Roy F, Mareel M, Birchmeier W. J Cell Biol 1993;120:757–766.
6. Matsuyoshi N, Hamaguchi M, Tanigushi S, Nagafuchi A, Tsukita S, Takeichi M. J Cell Biol 1993; 118:703–717.
7. Shiozaki H, Tahara H, Oka H, Miyata M, Kobayashi K, Tamura S, Iihara K, Doki Y, Hirano S, Takeichi M. Am J Pathol 1991;139:17–23.

8. Gamallo C, Palacios J, Suarez A, Pizarro A, Navarro P, Quintanilla M, Cano A. Am J Pathol 1993; 142:987–993.
9. Oka H, Shiozaki H, Kobayashi K, Inoue M, Tahara H, Kobayashi T, Takatsuka Y, Matsuyoshi N, Hirano S, Takeichi M, Mori T. Cancer Res 1993;53:1696–1701.
10. Vleminckx K, Vakaet L Jr, Mareel M, Fiers W, Van Roy F. Cell 1991;66:107–119.
11. Frixen UH, Behrens J, Sachs M, Eberle G, Voss B, Warda A, Löchner D, Birchmeier W. J Cell Biol 1991;113:173–185.
12. Chen W, Öbrink B. J Cell Biol 1991;114:319–327.
13. Behrens J, Mareel MM, Van Roy FM, Birchmeier W. J Cell Biol 1989;108:2435–2447.
14. Mareel MM, Behrens J, Birchmeier W, De Bruyne GK, Vleminckx K, Hoogewijs A, Fiers WC, Van Roy FM. Int J Cancer 1991;47:922–928.
15. Soule HD, Vazquez J, Long A, Albert S, Brennan M. J Natl Cancer Inst 1973;51:1409–1416.
16. Coopman P, Bracke M, Lissitzky J-C, De Bruyne GK, Van Roy FM, Foidart J-M, Mareel MM. J Cell Sci 1991;98:395–401.
17. Mareel M, Kint J, Meyvisch C. Virchows Arch B Cell Pathol 1979;30:95–111.
18. Bracke ME, Van Larebeke NA, Vyncke BM, Mareel MM. Br J Cancer 1991;63:867–872.
19. Bracke ME, Vyncke BM, Bruyneel EA, Vermeulen SJ, De Bruyne GK, Van Larebeke NA, Vleminckx K, Van Roy FM, Mareel MM. Br J Cancer 1993;68:282–289.
20. Kadmon G, Korvitz A, Altevogt P, Schachner M. J Cell Biol 1990;110:193–208.
21. Shimoyama Y, Hirohashi S, Hirano S, Nogushi M, Shimosato Y, Takeichi M, Abe O. Cancer Res 1989;49:2128–2133.
22. Plessers L, Bosmans E, Cox A, Op de Beek L, Vandepitte J, Vanvuchelen J, Raus J. Anticancer Res 1990;10:271–277.
23. Daughaday WH. In: Schofield PN (ed) The Insulin-Like Growth Factors. Oxford-New York-Tokyo: Oxford University Press, 1992;5–11.
24. Jennische E, Isgaard J, Isaksson OGP. In: Schofield PN (ed) The Insulin-like Growth Factors. Oxford-New York-Tokyo: Oxford University Press, 1992;221–239.
25. Sara VR. In: Schofield PN (ed) The Insulin-like Growth Factors. Oxford-New York-Tokyo: Oxford University Press, 1992;258–279.
26. De Leon DD, Bakker B, Wilson DM, Hintz RL, Rosenfeld RG. Biochem Biophys Res Commun 1988;152:390–405.
27. Mountjoy KG, Finlay GJ, Holdaway IM. Cancer Res 1987;47:6500–6504.
28. Moxham CP, Duronio V, Jacob S. J Biol Chem 1989;264:13238–13244.
29. Jacobs J, Kull FCJr, Earp HS, Svoboda ME, Van Wyck JJ, Cuatrecasas P. J Biol Chem 1983;258: 9581–9584.
30. Kadowaki T, Koyasu S, Nishida E, Tobe K, Izumi T, Takaku F, Sakai H, Yahara I, Kasuga M. J Biol Chem 1987;262:7342–7350.
31. Skolnik EY, Batzer A, Li N, Lee CH, Lowenstein E, Mohammadi M, Margolis B, Schlessinger J. Science 1993;260:1953–1955.
32. Baltensperger K, Kozma LM, Cherniack AD, Klarlund JK, Chawla A, Banerjee U, Czech MP. Science 1993;260:1950–1952.
33. Palka J, Peterkofsky B. Anal Biochem 1988;175:442–449.
34. Figueroa JA, Yee D. Breast Cancer Res Treat 1992;22:81–90.
35. Sara VR, Carlsson–Skwirut C, Andersson C, Hall E, Sjogren B, Holmgren A, Jornvall H. Proc Natl Acad Sci USA 1986;83:4904–4907.
36. Ogasawara M, Karey KP, Sirbasku DA. Proc AACR 29 1988;A207.
37. Cullen KJ, Smith HS, Hill S, Rosen N, Lippman ME. Cancer Res 1991;51:4978–4985.
38. Paik S. Breast Cancer Res Treat 1992;22:31–38.
39. Peyrat JP, Bonneterre J. Breast Cancer Res Treat 1992;22:59–67.
40. Papa V, Gliozzo B, Clark GM, McGuire WL, Moore D, Fujita-Yamaguchi Y, Vigneri R, Goldfine ID, Pezzino V. Cancer Res 1993;53:3736–3740.
41. Martin PM, Horwitz KB, Ryan DS, McGuire WL. Endocrinology 1978;103:1860–1867.

42. Akiyama T, Ishida J, Nakagawa S, Ogawara H, Watanabe S, Itoh N, Shibuya M, Fukami Y. J Biol Chem 1987;262:5592–5597.
43. Graziani Y. In: Cody V, Middleton E Jr, Harborne JB (eds) Plant Flavonoids in Biology and Medicine: Biochemical, Pharmacological and Structure-Activity Relationships. New York: Alan R. Liss Inc, 1986;301–313.
44. Bracke ME, Castronovo V, Van Cauwenberge RA-M, Coopman P, Vakaet L Jr, Strojny P, Foidart J-M, Mareel MM. Exp Cell Res 1987;173:193–205.
45. Ching LM, Baguley BC. Eur J Cancer Clin Oncol 1987;23:1047–1050.
46. Bracke ME, Vyncke BM, Van Larebeke NA, Bruyneel EA, De Bruyne GK, De Pestel GH, De Coster WJ, Espeel M, Mareel MM. Clin Exp Metastasis 1989;7:283–300.
47. Bracke M, Vyncke B, De Pestel H, Vakaet JrL, Bourgois L, Van Larebeke N, Bortier H, Mareel H. In: Das NP (ed) Plant Flavonoids in Biology and Medicine III. Current Issues in Flavonoid Research. Singapore: National University of Singapore Press, 1989;279–292.
48. Kopper L, Van Hahn T, Lapis K. J Cancer Res Clin Oncology 1983;103:31–38.
49. Giavazzi R, Garofalo A, Damia G, Garattini S, D'Incalci M. Br J Cancer 1988;57:277–280.
50. Bracke M, Romijn H, Vakaet L Jr, Vyncke B, De Mets M, Mareel M. In: Bjerkvig R (ed) Spheroid Culture in Cancer Research. Boca Raton-Ann Arbour-London: CRC Press, 1992;73–105.
51. Giguere V, Ong ES, Segui P, Evans RM. Nature 1987;330:624–629.
52. Eichele G. Ann NY Acad Sci 1993;678:22–36.
53. Mader S, Leroy P, Chen JY, Chambon P. J Biol Chem 1993;268:591–600.
54. Rijksen G, Voller MC, van Zoelen EJ. FEBS Lett 1993;322:83–87.

Variants of CD44 in human breast cancer

Karl-Heinz Heider[1], Hans-Peter Sinn[2], Gunther von Minckwitz[3], Manfred Kaufmann[3], Steven T. Pals[4], Peter Herrlich[1] and Helmut Ponta[1]

[1]*Kernforschungszentrum Karlsruhe, Institut für Genetik, Postfach 3640, D-76021 Karlsruhe, Germany;* [2]*Institut für Pathologie der Universität Heidelberg, Im Neuenheimer Feld 220, D-69120 Heidelberg, Germany;* [3]*Frauenklinik der Universität Heidelberg, Voßstr. 9, D-69115 Heidelberg, Germany; and* [4]*University of Amsterdam, Academic Medical Center, Department of Pathology, Meibergdreef 9, NL-1105 AZ Amsterdam, The Netherlands*

CD44 designates a polymorphic family of surface proteins, which are expressed on various cell types and under various conditions. The polymorphism is based on posttranslational modifications, mainly N-glycosylations, but also by changes in the primary amino acid structure due to splice variation [1]. The smallest isoform, CD44s (standard), is, at least in humans, derived from 10 different exons of which seven encode the N-terminal extracellular part of the protein. One accounts for the transmembrane region, and two code for a short (71aa) cytoplasmic tail [2]. Differential splicing of "variant" exons (nine in human and 10 in rat and mouse) in a multitude of combinations accounts for the occurrence of larger CD44 proteins with extended extracellular portions, commonly designated as CD44 variants (CD44v) [2,3]. In CD44s, these variant exons, which are located between exon five and six of the standard form, are absent.

CD44s is the most abundantly expressed isoform. It is expressed in many different cell types, particularly in hematopoietic cells, but also in dermal fibroblasts. The larger splice variants are far more restricted in their expression [4]. One cell type which expresses these variants are skin keratinocytes. Interestingly, variants of CD44 are also transiently expressed in lymphocytes upon antigenic stimulation, suggesting a crucial role of variants in the immune response [5,6]. Such a role was confirmed by the observation that specific antibodies recognizing the variant part of CD44 could abolish the immune response [6].

Our interest in variants of CD44 was based on the observation that they are expressed in metastasizing rat tumor cells, but not in related nonmetastasizing cells [7]. Upon subcutaneous injection of the metastasizing tumor cells together with CD44 variant-specific antibodies into syngeneic animals, we found that metastatic spread was retarded and in some instances even suppressed, indicating a causal role of CD44 variants in metastasis formation [8,9]. Furthermore, expression of CD44 variants upon transfection into nonmetastasizing tumor cells induced the metastatic phenotype in these cells [7,10].

A variety of different CD44 variants have since been transfected into locally growing tumor cells and many of them also induced metastatic behavior in the cells

[11]. Common to all these variants was the expression of the variant exon v6. The smallest CD44 variant with the ability to induce metastases tested so far, is one containing only exons v6 and v7. The standard type of CD44, however, never led to metastasis induction upon transfection into locally growing tumor cells and subsequent subcutaneous injection of these cells into syngeneic animals [7,11].

To unravel a possible role of CD44 variants in human cancer, we have cloned human variant CD44 cDNAs. In a first screen of human tumor cell lines using PCR techniques, we detected expression of CD44 variants. To extend our search to tumor material we raised a polyclonal serum and monoclonal antibodies against human variant CD44 sequences [4,5,12]. The specificity of these antibodies was checked by Western blot analysis, using single CD44 variant exons expressed as glutathione S-transferase fusion proteins in bacteria [4,5,12]. Their usefulness in detecting CD44 variants expressed in eucaryotic cells by immunohistochemical staining was tested on skin sections, because we know from PCR data that keratinocytes express a variant encoded by the exons v3 to v10. With the exception of exons v3 and v9, we isolated monoclonal antibodies to all human exons and they all react with human skin keratinocytes. Thus we could use these antibodies to examine CD44 variant expression on frozen sections of human tumors. We examined a variety of human tumors for CD44v expression. While for some tumors the expression of CD44v appeared to have no bearing on the status of the tumor, CD44v expression is clearly correlated with tumor progression and metastasis in several tumor types. The major results of these experiments are shown below.

We have preliminary data with a rather small sample number of skin-derived tumors. In six samples of melanomas we detected no expression of variants of CD44 containing v6 sequences on the surface. Thus, although melanomas are highly aggressive tumors, variants of CD44 containing this exon are not likely to be involved in tumor progression. In contrast, basaliomas are a type of tumor that never give rise to metastatic spread. We examined 11 samples of basaliomas and all of them showed expression of exon v6-containing CD44 variants. Since basaliomas are derived from skin keratinocytes, which are themselves positive for CD44 variants, this could indicate that upon transformation, expression of CD44 variants was maintained. Thus, in the case of basal cell carcinomas, the expression of CD44 v6-containing variants is not sufficient to mediate metastasis formation.

A first clue for the relevance of CD44 variant expression in human tumor progression came from examination of human non-Hodgkin's lymphoma samples. There was a direct correlation between tumor progression as defined by tumor grade and the expression of CD44 variants [5]. None of the low grade lymphomas expressed immunohistochemically detectable levels of CD44 variants, while 50% of the intermediate and high-grade lymphomas were positive for staining with CD44 v6-specific antibodies.

All gastric adenocarcinoma samples (42) revealed variant CD44 expression both at the RNA and surface protein level [12]. Interestingly, the two main types of gastric tumors, intestinal and diffuse, differ with respect to the type of CD44 variants expressed. The intestinal type tumors are positive for epitopes encoded by variant exons v5 and v6, whereas diffuse type carcinomas express only v5. In this case, the normal stomach mucosa is also positive for expression of exon v5 within the foveolar proliferation zone and on mucoid surface epithelium, while intestinal metaplasia reacts positively with antibodies specific for exons v5 and v6. This observation can be taken as an indication that the two different types of gastric tumors are derived from different types of normal cells. More importantly, however, it establishes variants of CD44 as a molecular parameter for diagnosis of gastric cancer. With respect to tumor progression, there was no correlation of expression of CD44 variants with more progressed stages of tumors. In particular, metastases derived from intestinal type gastric tumors were positive for CD44 exons v5 and v6 and metastases derived from diffuse type carcinomas expressed only exon v5.

Colorectal tumor progression is the best defined human cancer system with respect to pathological stages and the occurrence of molecular markers [13]. We have examined 70 samples of colorectal tumors of different stages and benign lesions for the expression of CD44 variants [5,14]. The expression of CD44 variants is directly linked with the transformation process since normal colorectal mucosa stains only very weakly at the base of the crypts with variant-specific antibodies (polyclonal serum directed against variant exons v3 − v10), whereas tumor samples are all positive for CD44 variant expression. Due to the fact that very early stages of colorectal carcinogenesis, e.g., early adenomas (diameter ≤1 cm), already express CD44 variants, the expression of CD44 variants is an early event. An interesting picture emerges if one uses exon specific monoclonal antibodies to define the type of CD44 variants expressed in these tumors. With v5- and v10-specific monoclonal antibodies, the result previously obtained with the polyclonal serum was confirmed. In contrast to this finding, expression of variant exon v6 appears to be strongly correlated with tumor progression. Normal epithelium is devoid of any CD44-v6 expression, as are early adenomas. In late adenomas (diameter >1 cm) we found about 40% positive tumors, and in most progressed stages of carcinomas (Dukes C/D), more than 90% of the tumors expressed CD44-v6 variants [14]. This increase of expression of CD44 variants containing exon v6 sequences might reflect a selective advantage of these tumor cells at later stages of tumor development and is compatible with the idea that such variants play an important role in the formation of metastases in humans.

A more conclusive interpretation of CD44v staining could be obtained if follow-up studies of patients whose tumors were tested showed a correlation of the expression of CD44 variants with the prognosis of survival. Such follow-up data have been collected for human breast cancer patients at the Frauenklinik in Heidelberg. One hundred and twenty-one tumor samples were chosen which were predominantly derived from primary tumors, ductal type as well as lobular type, and from lymph node metastases. Immunohistological staining with exon-specific antibodies revealed that about 75% of the primary tumors were positive for expression of v3- to v6-containing CD44 variants. Interestingly, all the lymph node metastases examined were positive for the expression of these exons, suggesting that the expression correlates with more progressed stages of cancer [15]. Normal ductal and lobular mammary epithelium and benign hypertrophies did not express these CD44 variants. However, expression was observed in the myoepithelial cell layer surrounding the ducts and lobuli.

To test whether these data have prognostic value, they were combined with the follow-up data of the patients [Kaufmann M, Heider K-H, Sinn H-P, von Minckwitz G, Ponta H, Herrlich P, submitted]. In these follow-up studies, other data including lymph node status, ploidy, S-phase fraction, c-erb B2 and EGF-R expression were also available. Univariate analysis shows that positive v3 to v6 epitope staining is a significant predictor of reduced overall survival. In this collection of patients the predictive capacity of the CD44 variant expression appears to be better than that of the nodal status and S-phase fraction and similar to histological grading. These data suggest that the expression of CD44 variants is an even better predictor of overall survival of patients than lymph node status, which is the clinically most important prognostic factor to date [16,17].

From our data, CD44 variants containing exon sequences v3 to v6 appear to confer selective advantage during tumor progression, e.g., by conferring a growth advantage and/or metastatic properties to human breast cancer cells. The strong correlation of CD44 variant expression with poor survival suggests a causal role of CD44 variants in the formation of distant metastases in human breast cancer. Thus, the identification of a causal role of CD44 variant expression in the metastatic spread of tumor cells originally established in an animal system, appears also to hold true for human breast cancer, and possibly other tumors.

Acknowledgements

We thank Ingrid Kammerer and Diane Nichol for excellent secretarial assistance, and Jonathan Sleeman for critically reading the manuscript. The work was supported by a grant from the Deutsche Forschungsgemeinschaft.

References

1. Haynes et al. Cancer Cells 1991;3:347–350.
2. Screaton GR, Bell MV, Jackson DG, Cornelis FB, Gerth U, Bell JI. Proc Natl Acad Sci USA 1992; 89:12160–12164.
3. Tölg C, Hofmann M, Herrlich P, Ponta H. Nucl Acid Res 1993;21:1225–1229.
4. Heider K-H, Hofmann M, Horst E, van den Berg F, Ponta H, Herrlich P, Pals ST. J Cell Biol 1993; 120:227–233.
5. Koopman G, Heider K-H, Horst E, Adolf GR, van den Berg F, Ponta H, Herrlich P, Pals ST. J Exp Med 1993;177:897–904.
6. Arch R, Wirth K, Hofmann M, Ponta H, Matzku S, Herrlich P, Zöller M. Science 1992;257:682–685.
7. Günthert U, Hofmann M, Rudy W, Reber S, Zöller M, Haußmann I, Matzku S, Wenzel A, Ponta H, Herrlich P. Cell 1991;65:13–24.
8. Seiter S, Arch R, Komitowski D, Hofmann M, Ponta H, Herrlich P, Matzku S, Zöller M. J Exp Med 1993;177:443–455.

9. Reber S, Matzku S, Günthert U, Ponta H, Herrlich P, Zöller M. Int J Cancer 1990;46:919–927.
10. Rudy W, Hofmann M, Schwartz-Albiez R, Zöller M, Heider K-H, Ponta H, Herrlich P. Cancer Res 1993;53:1262–1268.
11. Herrlich P, Rudy P, Hofmann M, Arch R, Zöller M, Zawadzki V, Tölg C, Hekele A, Koopman G, Pals S, Heider K-H, Sleeman J, Ponta H. In: Hemler ME and Mihich E (eds) Cell Adhesion Molecules. New York and London: Plenum Press, 1993;265–288.
12. Heider K-H, Dämmrich J, Skroch-Angel P, Müller-Hermelink H-K, Vollmers HP, Herrlich P, Ponta H. Cancer Res 1993;53:4197–4203.
13. Fearon ER, Vogelstein B. Cell 1990;61:759–767.
14. Wielenga V, Heider K-H, Offerhaus JG, Adolph GR, van den Berg F, Ponta H, Herrlich P, Pals ST. Cancer Res 1993;53:4754–4756.
15. Sinn H-P, Heider K-H, Skroch-Angel P, von Minckwitz G, Kaufmann M, Herrlich P, Ponta H. Human mammary carcinomas express homologues of metastasis-associated variants of CD44. Breast Cancer Res Treat (in press).
16. Clark GM, McGuire WL, Hubay CA, Pearson OH, Marshall JS. N Engl J Med 1983;309:1343–1347.
17. Glick JH. J Nat Cancer Inst 1992;84:1479.

Induction by urokinase-type plasminogen activator of c-*fos* gene expression in human ovarian cancer cells

I. Dumler[*], T. Petri[*] and W.-D. Schleuning[*]

Research Laboratories of Schering AG, 13342 Berlin, Germany

Abstract. Binding of urokinase-type plasminogen activator (u-PA) to u-PA receptor (u-PAR) induces the rapid and transient expression of c-*fos* in OC-7 ovarian carcinoma cells. The pretreatment of the cells with protein tyrosine kinase (PTK) inhibitors, but not the inactivation of the u-PA active site by diisopropyl fluorophosphate (DFP), abrogates this effect. A soluble u-PAR fragment, expressed in baculovirus infected Sf9 cells and purified by affinity chromatography, competes for binding of u-PA to u-PAR and inhibits c-*fos* induction. It is concluded that activation of u-PAR after interaction with u-PA at the cell surface initiates a transmembrane signal, most likely in conjunction with other still unknown protein(s). This signal generates PTK activity feeding into a signal transduction pathway which activates nuclear transcription factors.

Introduction

Invasion and metastasis of solid tumors require the action of tumor-associated proteases which promote the dissolution of the surrounding cellular matrix and basement membranes [1]. Urokinase-type plasminogen activator (u-PA) appears to play an important role in these events [2]. In addition to its proteolytic activity, u-PA interacts as a ligand with its specific receptor. Whereas it is generally held that the receptor focuses u-PA activity in space and time [3], recent data indicates that u-PA receptor (u-PAR), after the binding of u-PA, might interact with other transmembrane proteins in order to transduce a signal [4]. We have shown recently that in histiocytic lymphoma cells (U937), u-PA acts as an effector by activating a protein tyrosine kinase [5]. The distal components of this signalling pathway, which might regulate cell proliferation, are unknown. Likely candidates for such a role could be products of immediate early genes, which are essential for the stimulation of cell proliferation by growth factors [6]. Therefore we investigated the effects of u-PA on c-*fos* gene expression in OC-7 human ovarian cancer cells, and report here that activation of u-PAR by u-PA induces a rapid and transient expression of c-*fos* which is mediated via a tyrosine kinase.

[*]*Address for correspondence:* I. Dumler, T. Petri and W.-D. Schleuning. Institute of Cellular and Molecular Biology, Schering AG, Mullerstr. 170, 13342 Berlin, Germany. Tel.: +49-30-468-1390. Fax: +49-30-46916707.

Experimental

Materials

Chemicals were of the best commercial grade available and purchased from Sigma (St Louis, MO), Pharmacia LKB Biotechnology (Uppsala, Sweden), Merck (Darmstadt, Germany) or Serva (Heidelberg, Germany). Radiochemicals were obtained from Amersham International (Little Chalfont, Great Britain), PI-specific PLC was from Sigma, the protein tyrosine kinase (PTK) inhibitors Herbimycin A (*Streptomyces spp.*) and erbstatin analogue (methyl2,5-dihydroxycinnamate) were from Calbiochem Biochemicals (San Diego, USA), u-PA was from Serono (Freiburg, Germany), anti-u-PAR monoclonal antibody (product #3936) was purchased from American Diagnostica Inc. (Greenwich, USA) and purified rabbit antiphosphotyrosine polyclonal antibodies was from Dianova (Hamburg, Germany). The mRNA purification kit was obtained from Pharmacia LKB Biotechnology (Uppsala, Sweden) and the multiple DNA labelling system from DuPont NEN (Boston, USA). All cell culture media and the fetal calf serum (FCS) were provided by Gibco (Karlstein, Germany).

Cell culture

OC-7 cells, a cell line isolated from a human cystadenocarcinoma, were provided by Prof M. Schmitt (Technical University Munich, Germany) and grown in DMEM/ HAM F12 nutrient mix containing 5% FCS.

Sf9 insect cells were received from the European Collection of Animal Cell Cultures (ECACC, Porton Down, Wilts, UK.) and either propagated in supplemented Grace Insect (TNMFH) medium containing 10% FCS and 0.1% pluronic (Serva, Heidelberg, Germany) or adapted to growth in serum free medium (Sf 900 II medium) in this laboratory.

Electrophoresis, Western blotting and autoradiography

Sodium dodecyl sulphate-polyacrylamide gel electrophoresis (SDS-PAGE) was carried out in slab gels (7.5, 10 or 12.5%) as described by Laemmli [7]. Samples were reduced immediately before electrophoresis in the presence of 20 mM dithiothreitol (DTT) for 5 min at 95°C or analyzed under nonreducing conditions. Gels were electroblotted onto nitrocellulose sheets which were subsequently blocked with 1% bovine serum albumin (BSA) or 30% FCS [5]. Autoradiography of radiolabelled proteins was performed at 70°C with dried polyacrylamide gels using Konica X-ray film.

Protein determination

Protein was quantified using a bicinchoninic acid (BCA) reagent from Pierce, using BSA as a standard.

Northern blot analysis

Total RNA was isolated from 5×10^7 cells by the guanidinium isothiocyanate method as described by Sambrook et al. [8]. In some experiments polyadenylated RNA was purified from total RNA on oligo (dT)-cellulose columns using an mRNA purification kit. 20 µg of total RNA or 5 µg of polyadenylated RNA were separated on 1% agarose — 18% formaldehyde gels and transferred onto Gene Screen Plus sheets by capillary force. A fragment of the c-*fos* gene (1 kb) was labelled with ^{32}P-dCTP by random-priming [9]. The membranes were prehybridised in a solution containing 50% formamide, 1% SDS, 10% dextranesulfate and 1 M NaCl at 42°C for 6 h and then hybridised with a labelled probe (1×106 cpm/ml) in the presence of denatured salmon sperm DNA (100 µg/ml) at 42°C overnight. The filters were washed twice with 2×0.15 M NaCl/0.015 M Na_3-citrate, pH 7.0 (SSC) (each time for 5 min at room temperature), followed by another two washes with $2 \times$ SSC/1% sodium dodecyl sulphate (SDS) (each time for 30 min at 60°C). Finally the filters were dried at room temperature and exposed to Konica X-ray film at -70°C.

Construction of a recombinant baculovirus, expressing soluble u-PAR

The complete cDNA of the human urokinase-type plasminogen activator receptor (u-PAR) was obtained from Dr E.K.O. Kruithof (Centre Hospitalier Universitaire Vaudois, Lausanne, Switzerland). Using synthetic primers corresponding to the published u-PAR cDNA sequence [10], a 922 base pair PCR fragment was amplified corresponding to the amino terminal 285 amino acids of the receptor, lacking the site for attachment of the GPI anchor. The primers were designed so that a 5'-end *Eco* RI site and a 3'-end *Bgl* II site were generated at each end of the fragment. Via these two restriction sites, the truncated u-PAR cDNA was inserted into the polylinker site of the baculovirus transfer vector pVL1393 (InVitrogen, San Diego, CA USA) downstream of the polyhedrin promoter.

A recombinant baculovirus was then constructed by cotransfection of Sf9 insect cells with the transfer vector DNA together with linearized viral DNA ("Baculogold virus DNA" from Pharmingen, Palo Alto, CA, USA). Transfection and isolation of the recombinant virus was performed as described by King et al. [11]. Soluble u-PAR was produced by infection of Sf9 cells adapted to growth in serum-free medium (Sf 900 II medium; Gibco) with the recombinant virus at a multiplicity of infection of five. The culture supernatant from infected cells containing the secreted soluble receptor was harvested 60 h post infection.

Purification of recombinant soluble u-PAR

u-PAR was purified from culture supernatants by affinity chromatography on u-PA-Sepharose which was prepared as previously described by Dumler et al. [5]. The supernatant containing the expressed u-PAR was passed over u-PA-Sepharose in the presence of 1 M NaCl. Bound u-PAR was eluted from the column with 0.2 M acetic

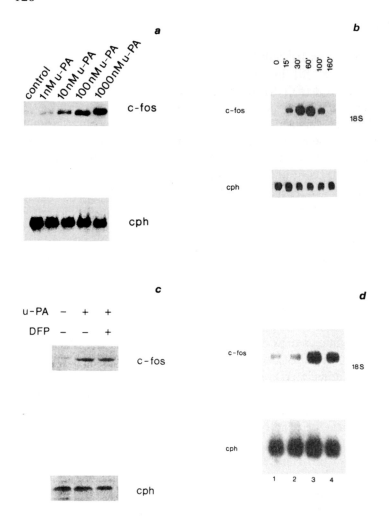

Fig. 1. u-PA-induced c-*fos* gene expression in OC-7 cells.
(a) The cells were stimulated with different concentrations of u-PA for 30 min. Total RNA (20 µg) was analyzed by Northern blot hybridization, as described in Materials and Methods. Human cyclophilin mRNA levels were assessed as a control; (b) The cells were incubated with 100 nM u-PA for 0–160 min; (c) u-PA was inactivated by the treatment with 20 mM DFP for 2 h at room temperature (right lane); (d) OC-7 cells were pretreated with herbimycin A (1 µM/2 × 10^5 cells) for 1 h at 37°C before u-PA stimulation (lane 2); lane 1 – control; lane 3 – stimulation with u-PA (100 nM, 30 min); lane 4 – stimulation with PMA (150 nM) for 2 h.

acid adjusted to pH 1.8 by the addition of HCl and immediately neutralized with Tris to pH 8.0. Before the assay, u-PAR was dialyzed against phosphate buffered saline (PBS) and concentrated using polyethylene glycol (M_r = 20,000).

Amino-terminal amino acid sequence analysis of u-PAR was performed after electrophoretic transfer of the protein from polyacrylamide to Immobilon membranes.

Results and Discussion

u-PA-induced c-fos gene expression

Northern blot analysis demonstrated that u-PA-induced c-*fos* expression was dose-dependent (Fig. 1a) and that the increase of mRNA started at physiological concentrations of the ligand.

Of additional interest was the time-course of the induction, in comparison to other effectors [12]. As shown in Fig. 1b, the mRNA level reached a maximum after 30 min treatment and declined to undetectable levels after 120 min. This type of kinetics resembles the one previously described in the context of mitogen-induced c-*fos* expression [13], but not the type observed after induction with muscarinic receptor agonists and platelet-activating factor, where transcription started within 1 min and detectable signals were observed as early as 5 min after ligand addition [14].

In order to exclude a potential contribution of proteolytic activity to the mechanism of induction, u-PA was inactivated with diisopropyl fluorophosphate (DFP) (Fig. 1c). This material induced the same level of expression, corroborating previous findings that demonstrated the involvement of proteolytically inactive u-PA in signalling events [4,5]. Stimulation of tyrosine kinases is a common denominator of various signalling pathways (including signalling via GPI-anchored proteins, [15]), ultimately leading to the activation of nuclear transcription factors [16]. Pretreatment of cells for 20 min with Herbimycin A, an inhibitor of protein tyrosine kinase activity, completely inhibited u-PA-induced signal transduction (Fig. 1d). The same results were obtained with the erbstatin analogue (data not shown). Phorbol 12-myristate acetate (PMA)-treated cells served as a positive control.

Expression and purification of recombinant truncated u-PAR

Recombinant truncated u-PAR, purified by affinity chromatography on u-PA — Sepharose, was subjected to SDS-PAGE and Western blot analysis. Under nonreducing conditions one prominent band ($M_r \sim 35$ kDa) was stained by Coomassie Blue R-250 (Fig. 2a) and revealed by Western blot analysis using polyclonal and monoclonal anti-u-PAR antibodies. The amino acid sequence of purified Sf9-u-PAR was determined by automated Edmanndegradation. The reduced M_r of the fragment indicated incomplete glycosylation during heterologous expression. These data established unequivocally that the purified material consisted of authentic, truncated u-PAR.

Competition for binding by soluble u-PAR

To characterize the functional activity of recombinant u-PAR, a competition assay was performed. u-PA was preincubated with increasing concentrations of recombinant Sf9-u-PAR and then incubated with OC-7 cells for 30 min. The induction of c-*fos* expression in all probes was evaluated by Northern blot analysis. The data presented

128

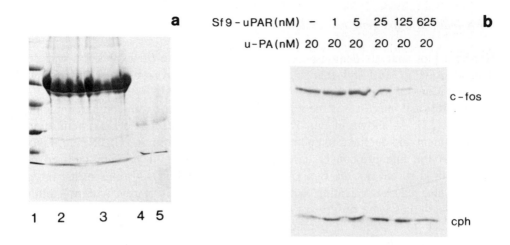

Fig. 2. (a) Purification of recombinant Sf9-u-PAR by affinity chromatography. Details of the purification are described in the text. Lane 1 — molecular weight markers; lanes 2 and 3 — culture supernatant; lanes 4 and 5 — fractions after affinity chromatography; (b) u-PA (20 nM) was preincubated with increasing concentrations of recombinant u-PAR and then incubated with the cells as described in the text. RNA extraction and Northern blot analysis were carried out as described in Materials and Methods.

on Fig. 2b clearly shows that recombinant u-PAR competed with binding of u-PA to u-PAR and interfered with u-PA-induced c-*fos* expression in a dose-dependent fashion. A 30-fold molar excess of recombinant Sf9-u-PAR almost completely prevented the induction.

Our results allow the following conclusions: 1) u-PA transduces signals to the nuclear transcriptional apparatus via u-PAR. This effect most likely requires the interaction of u-PAR with still undefined additional protein(s); 2) The signal is generated independently of u-PA proteolytic activity; and 3) Involves the activation of a tyrosine kinase. What we described represents probably a novel pathway of growth control that might be accessible to therapeutic intervention.

Acknowledgements

We would like to thank Prof M. Schmitt for providing OC-7 cell line, Prof F. Blasi for anti-u-PAR monoclonal antibodies, Dr E.K.O. Kruithof for complete cDNA of human u-PAR and Dr A.C.B. Cato for the c-*fos* DNA probe. The technical assistance of D. Schmidt and O. Lange was greatly appreciated.

References

1. Alitalo K, Vaheri A. Adv Cancer Res 1982;37:111—158.
2. Danø K, Anderson PA, Grøndahl-Hansen J, Kristensen P, Nielsen LS, Skriver L. Adv Cancer Res

1985;44:139–266.

3. Kristensen P, Eriksen J, Blasi F, Danø K. J Cell Biol 1991;115:1763–1771.

4. Odekon LE, Sato Y, Rifkin DB. J Cell Physiol 1992;150:258–263.

5. Dumler I, Petri T, Schleuning W-D. FEBS Lett 1993;322:37–40.

6. Distel RJ, Spiegelman BM. Adv Cancer Res 1990;55:37–55.

7. Laemmli UK. Nature 1970;227:680–685.

8. Sambrook J, Fritsch EF, Maniatis T. In: Molecular Cloning. A Laboratory Manual, 2nd ed. Cold Spring Harbor, New York: Cold Spring Harbor Laboratory Press, 1989.

9. Feinberg AP, Vogelstein B. Anal Biochem 1983;132:6–13.

10. Roldan AL, Cubellis MV, Masucci MT, Behrendt N, Lund LR, Danø K, Appella E, Blasi F. EMBO J 1990;9:467–474.

11. King LA, Possee RD. In: The Baculovirus Expression System. A Laboratory Guide. London, New York, Tokyo, Melbourne, Madras: Chapman & Hall Press, 1992.

12. Molinar-Rode R, Smeyne RJ, Curran T, Morgan JI. Mol Cell Biol 1993;13:3213–3220.

13. Herschman HR. Ann Rev Biochem 1990;60:281–319.

14. Ransone LJ, Verma JM. Ann Rev Cell Biol 1990;6:539–557.

15. Stefanova I, Horejsi V, Ansotegui IJ, Knapp W, Stockinger H. Science 1991;254:1016–1019.

16. Tripathi YB, Lim RW, Fernandez-Gallardo S, Kandala JC, Guntaka RV, Schukla SD. Biochem J 1992;286:527–533.

Prognostic factors,
new aspects of treatment

Synthetic peptides for the inhibition of cell-cell and cell-matrix adhesion

Roland Haubner[1], Marion Gurrath[2], Gerhard Müller[3], Monique Aumailley[4] and Horst Kessler[1*]

[1]*Institute of Organic Chemistry and Biochemistry, Technical University München, Germany; [2]University of Padua, Padua, Italy; [3]Glaxo, Verona, Italy; [4]C.N.R.S., University of Lyon, Lyon, France*

Abstract. Cell-surface receptors, such as the integrins, play an important role in metastasis. The integrins recognize the Arg-Gly-Asp (RGD) sequence in many extracellular proteins.

The conformationally controlled design of homodetic, cyclic penta- and hexapeptides together with biological assays provides an insight into the steric factors required for binding to the receptor. Two cyclic pentapeptides inhibit the laminin P1-fragment and vitronectin mediated tumor cell adhesion as well as the in vitro binding of fibrinogen to the isolated $\alpha_v\beta_3$ receptor.

Introduction

Metastasis, the spread of tumor cells from primary to distant, multiple secondary sites, is the major cause of morbidity and death for cancer patients [1,2]. The first step in this complex process leading to the formation of metastasis is the loss of cell-cell adhesion and the dissociation of individual cells from the primary tumor. After local invasion of adjacent tissue barriers, the tumor cell must invade the lymphatic system or directly penetrate the walls of the blood vessels in order to disseminate. Circulating tumor cells must survive the mechanical trauma of the blood flow, must be able to evade the attacks of the immune system and adhere in the venous or capillary bed of the target organ. The arrested tumor cells must again penetrate the vascular wall to enter the organ parenchyma. The extravasated tumor cell must be able to grow in foreign tissue different from that of the primary tumor in order to initiate a metastatic colony (Fig. 1) [1—4].

From a biochemical point of view, this process of metastasis involves alterations in homotypic cell-cell adhesion, cell migration and matrix degradations as well as heterotypic cell interactions during arrest and extravasation. On the one hand, they must be able to loose their cell-cell and/or cell-matrix adhesion contributing to initial release on invasion. On the other hand, they must possess the capability to form new adhesive properties related to the various steps of invasion and to arrest during

---

Address for correspondence: H. Kessler, Institute of Organic Chemistry and Biochemistry, TU München, Lichtenbergstraße 4, D-85747 Garching, Germany.

134

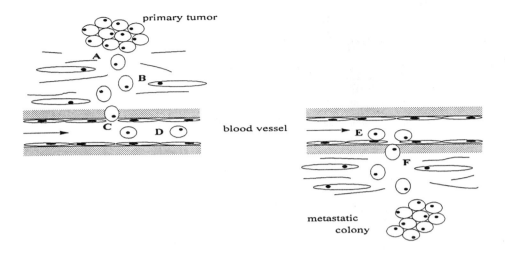

Fig. 1. Metastatic Cascade. A) Dissociation; B) Invasion of adjacent tissue; C) Invasion of the blood vessel; D) Circulation in the blood vessel; E) Arrest in the venous or capillary bed; F) Extravasation.

metastasis [3]. The cellular receptors mediating these adhesive events are thus likely to be important in tumor invasion and metastasis.

There are four major cell surface receptor classes which participate in cell adhesion and migration. Besides the cadherins (calcium-dependent adhesion molecules mediating homotypic cell-cell interactions), the immunoglobulin super-family (involved in many cell-cell and cell-substrate interactions both homotypic and heterotypic) and the selectins (mediating some heterotypic interactions between or among blood cells and endothelial cells), the integrins play an important role in adhesion during the metastatic cascade [3,5,6]. Each integrin is a heterodimeric glycoprotein consisting of an α-subunit (about 130–210 kDa) and a smaller β-subunit (about 95–130 kDa). Both subunits have plasma membrane-spanning sequences linking the internal cytoskeletal network of a cell with the external extracellular matrix. The cytoplasmic region of both units is typically short compared with the extracellular units. The extracellular domain of the β-subunit includes a highly conserved cysteine-rich motif consisting of four repeats close to the membrane-spanning region. The α-subunit is characterized by three or four stretches of about 12–15 amino acids, similar to calcium-binding motifs in other proteins like calmodulin [7] (Fig. 2). The specificity for ligand binding is determined by a particular combination of α- and β-subunits [8–11].

The integrins are divided into at least five subfamilies, each being defined by a common β-subunit. The most populated subfamilies are integrins with the β_1-, β_2- or β_3-subunit. The β_1 subfamily includes receptors for laminin, fibronectin and collagen, all of them being localized in the basement membrane. The β_2 subfamily is found on leukocytes and includes receptors which mediate cell-cell interactions. The well known platelet glycoprotein IIb/IIIa ($\alpha_{IIb}\beta_3$) and the vitronectin receptor $\alpha_v\beta_3$ belong to the β_3 subfamily [3].

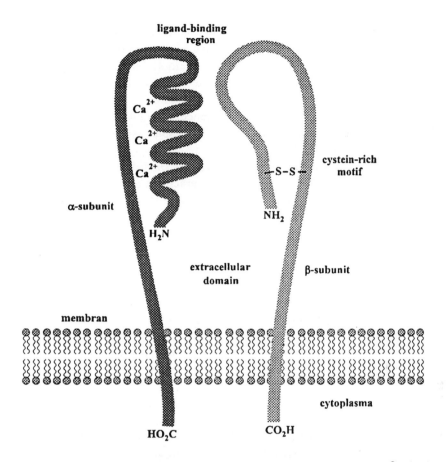

Fig. 2. Schematic presentation of a typical integrin consisting of an α- and a β-subunit.

It was demonstrated that there is a striking difference in the expression of the β_3-subunit between tumorigenic and nontumorigenic lesions, when the expression of integrins on cells embedded in tissue sections is examined. All cells from nevi, radial growth phase, vertical growth phase and metastatic lesions express the α_v-subunit. In contrast to this, only tumorigenic vertical growth phase and metastatic melanoma cells express the β_3-subunit, whilst benign melanocytic nevus and nontumorigenic radial growth phase primary melanoma uniformly failed to express this subunit [12]. Besides the vitronectin receptor, there is a variety of other integrins reported to be expressed on tumor cells [13–15].

It turned out that metastasis of several tumor cell lines can be inhibited by antibodies [13,16] or small, synthetic peptides [16–18] acting as ligands for these receptors. The tripeptide-sequence Arg-Gly-Asp (RGD in the single letter amino acid code) was found to be a common sequence of these peptides. The RGD-sequence, which is present in many extracellular matrix-proteins like vitronectin, laminin, fibronectin and fibrinogen, is a ligand recognition motif for several integrins. Despite this common motif, a high degree of specificity among the integrins is observed. This could be explained by assuming that the RGD-sequence serves as a shared binding site. Substrate-binding is mediated by the RGD-sequence, whereas the specificity for the different protein-ligands is determined by a second remote binding site. Another explanation is that the specificity arises from unique conformations of the RGD tripeptide in different matrix proteins. This active conformation is determined by the remainder of the protein. There is sufficient evidence accounting for the latter possibility, namely that the conformation of the RGD-sequence is the main factor in determining the ligand binding [8—11].

Therefore structural studies of integrins and their ligands should provide for deeper insights into the different cell-cell and cell-matrix interactions and their effects on tumor cell behavior. However, neither the structure of one of the integrins nor of receptor-ligand complexes are known. The structures of some proteins containing a biologically relevant RGD-sequence (Foot and Mouth Disease Virus [19,20], γ-Crystalline [21], Tenascin [22] and the tenth type III module of Fibronectin [23,24] as well as the snake venoms Kistrin [25,26], Echistatin [27—31] and Flavoridin [32]) have been determined, but in each structure the tripeptide sequence is exposed in the tip of a flexible loop or, in the case of the γ-Crystalline, in an extended edge-strand of a β-sheet. The flexible conformation of the RGD-motif allows recognition by different integrin-receptors but at the same time prohibits the determination of the bioactive conformation, necessary for a structure-based rational drug design.

Design

Our approach is based on the design of conformationally constrained peptide molecules [33]. Peptides are usually rather flexible entities and their conformations in solution may not necessarily correlate with the receptor-bound conformation. It is further known that the multiple interactions of peptides could arise from selection or induction of different conformers depending on the receptor environments [34]. Introduction of conformational constraints into a flexible peptide molecule can be used to develop models for the conformation required for receptor binding and activity [35]. Restriction in conformational freedom increases binding to a receptor if the biologically active conformation is included in the allowed conformational space. The reason is that less conformational entropy is lost upon binding and any strain necessary to adopt the binding conformation is preinduced. A drastic reduction of conformational space in peptide molecules is achieved by backbone cyclization. It also facilitates a conformational analysis by NMR spectroscopy and MD/DG-simulation methods [33,36,37].

If conformational restriction leads to inactive peptides, their three-dimensional structures can be used as a basis to define conformational elements which are not compatible with the binding site. Therefore the structure of an inactive peptide is as valuable as that derived from highly active ligands. In this sense the structure determination of active, conformationally restricted peptides allows the definition of

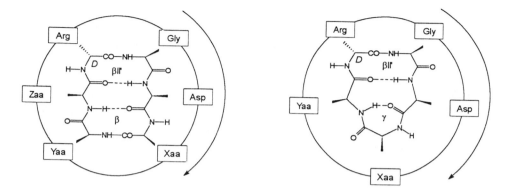

Fig. 3. Schematic representation of the structural templates for design of cyclic RGD-containing hexapeptide (left) and pentapeptide (right) analogues.The cyclic peptides provide scaffolds fixing the side-chains in a specific position. The amino acid sequence is shifted systematically around the template by using consecutively one amino acid in a given sequence in D-configuration. This amino acid occupies the upper left corner of the template.

the bioactive conformation [35]. Another advantage of such small, constrained peptides is their anticipated lower immunogenicity compared to larger peptides or proteins and their enhanced resistance against proteolysis that make them useful for therapeutic applications.

In our design the common RGD sequence has been exposed in well defined spatial arrangements. Conformational control is facilitated by incorporating D-amino acid residues and/or proline in the cyclic penta- or hexapeptide sequences.

A D-residue or proline induces characteristic turn motifs within the cyclic peptides. In both the pentapeptides and hexapeptides the D-amino acid strongly prefers the $i+1$ position of a βII′ turn. Generally, in cyclic hexapeptides a βII′ and a second β turn is formed, while in cyclic pentapeptides the D-residue generates a βII′/γ or a βII′/γ$_i$ turn arrangement [38]. Depending on the position of the D-residue within the RGD peptides, the essential tripeptide moiety is shifted around the common underlying turn templates (Fig. 3).

Thus, we induce different pharmacophore geometries of the primary sequence RGD by controlled utilization of pentapeptide and hexapeptide scaffolds. Following these design principles, we have synthesized different classes of cyclic pentapeptides and hexapeptides as shown in Figs. 4 and 5 [39]. Each of them has a different conformation regarding the RGD-sequence.

Class 1 shows hexapeptides with Arg in $i+1$ position of a βII′ turn. Shifting the D-amino acid counterclockwise by one position results in hexapeptides with Arg in $i+2$ of a βII′ turn. Further shifting of the D-amino acid leads to the different, conformational classes (Fig. 3–5).

In class 4 the essential RGD residues are in a comparable turn arrangement as in class 1, but with natural L-chirality of the RGD-sequence. Class 5 and 6 show two further structural families, but their turn motifs are already existing in classes 2 and 3 because of the pseudo-symmetry present in cyclic hexapeptides. The disadvantage

138

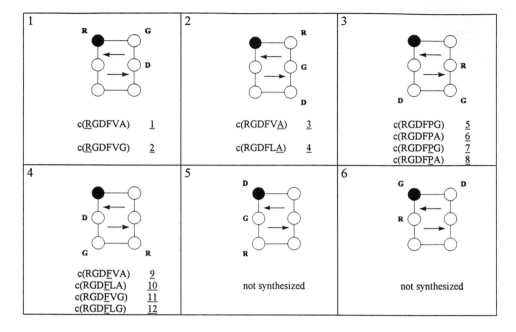

1	2	3
c(RGDFVA) 1 c(RGDFVG) 2	c(RGDFVA) 3 c(RGDFLA) 4	c(RGDFPG) 5 c(RGDFPA) 6 c(RGDFPG) 7 c(RGDFPA) 8
4	5	6
c(RGDFVA) 9 c(RGDFLA) 10 c(RGDFVG) 11 c(RGDFLG) 12	not synthesized	not synthesized

Fig. 4. Synthetic hexapeptide classes. The hexapeptide template is shown schematically with L-amino acids as plain circles and D-residues as filled circles. Arrows indicate intramolecular hydrogen bonds. The hydrogen bond pattern given here is based on extensive NMR measurements and MD calculations.

of class 5 is that the chirality of Asp would have to be changed to the unnatural D-form [39]. This modification leads to a dramatic loss in activity in linear and other cyclic peptides [40–42]. In order to synthesize peptides belonging to class 6, the essential Gly residue would have to be changed to a side-chain-bearing amino acid with D-chirality. The position within the RGD-sequence is very sensitive towards modification of the Gly residue. Introduction of a methyl group via alanine leads to decreased activities. To introduce further constraints into the RGD-sequence a variety of cyclic pentapeptides were synthesized following the same design principles as explained for the hexapeptides (Fig. 5) [for further details see Ref. 39].

Modern two-dimensional NMR-techniques and MD-simulation methods confirmed the conformations expected from our design principles [39].

Biological data

To investigate the structure/activity relationships of these penta- and hexapeptides the inhibition of the laminin P1-fragment and the vitronectin mediated cell adhesion of three different human cell lines (fibrosarcoma HT 1080, melanoma A 375 and mammary epithelia HBL-100) were examined [39,43]. In addition, the inhibition of the fibrinogen adhesion of two isolated integrin receptors (platelet receptor GP IIb/IIIa

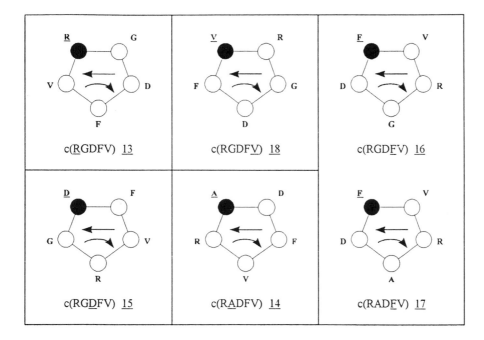

Fig. 5. Synthetic pentapeptides **13–18**. The pentapeptide template is shown schematically with L-amino acids as plain circles and D-residues as filled circles. Arrows indicate intramolecular hydrogen bonds. The hydrogen bond pattern given here is based on extensive NMR measurements and MD calculations.

and the vitronectin receptor) were investigated [44,45].

In the cell adhesion tests all of the hexapeptides were less inhibitory for cell adhesion to vitronectin when compared to the linear standard peptide GRGDS (Table 1). A more differentiated inhibitory profile was observed for the hexapeptides in cell adhesion with the laminin P1-fragment. The highest inhibitory activities were found for class 3 peptides (in the range of GRGDS), especially with HT 1080 and HBL-100 cell lines. All other hexapeptide classes show decreased inhibitory activities when compared to class 3 peptides (Table 2).

The inhibitory capacity of most of the hexapeptides towards fibrinogen adhesion of the isolated receptor is somewhat decreased compared to the linear standard peptide GRGDSPK. Only the peptides **4**, **11** and **12** are in the range of the standard peptide. Regarding the inhibitory profile of the hexapeptides concerning the inhibition of binding of fibrinogen to the isolated $\alpha_{IIb}\beta_3$ receptor [45], there is a more differentiated picture [44].

Similar to the cyclic hexapeptides, most of the cyclic pentapeptides were less active than the linear peptide GRGDS concerning the laminin P1-fragment and vitronectin mediated cell adhesion of the different cell lines. Only two peptides (**16** and **18**) showed significantly enhanced inhibitory activities. Whereas the D-Phe containing peptide **16** is active towards both the laminin P1-fragment and the vitronectin mediated cell adhesion, D-Val containing peptide **18** is selectively active

Table 1. Inhibitory capacity (IC$_{50}$) of cyclic penta- and hexapeptides towards cell adhesion to vitronectin [39]

Peptide[a]		IC$_{50}$ towards		
		HT 1080 (µM)	A 375 (µM)	HBL-100 (µM)
c(RGDFVA)	1	>120	>120	>120
c(RGDFVG)	2	>120	>120	>120
c(RGDFVA)	3	>120	>120	>120
c(RGDFLA)	4	120	>120	120
c(RGDFPG)	5	>100	>100	>100
c(RGDFPA)	6	>100	>100	>100
c(RGDFPG)	7	>100	>100	>100
c(RGDFPA)	8	>100	>100	>100
c(RGDFVA)	9	>120	67	124
c(RGDFLA)	10	118	120	28
c(RGDFVG)	11	>120	82	>120
c(RGDFLG)	12	115	78	>120
c(RGDFV)	13	>120	>120	>120
c(RADFV)	14	>120	>120	>120
c(RGDFV)	15	>120	>120	>120
c(RGDFV)	16	4	0.2	0.1
c(RADFV)	17	>120	46	41
c(RGDFV)	18	>120	20	30
RGDFV	19	>150	82	>170
GRGDS		>100	20	18

[a]The amino acid abbreviations are given in the IUPAC one letter code. For D-amino acids bold letters are used. R = Arg, G = Gly, D = Asp, F = Phe, V = Val, A = Ala, S = Ser, P = Pro.

to the laminin P1-fragment mediated cell adhesion (Table 1 and 2).

The profile of the pentapeptides towards inhibition of fibrinogen binding to the isolated $\alpha_{IIb}\beta_3$ and $\alpha_v\beta_3$ receptor [45] reveals similarities. Only the cyclic peptides **16** and **18** show increased activities compared to the standard peptide GRGDSPK [44].

Conclusions

The high activities of some cyclic peptides, together with the decreased activity of the linear RGDFV-peptide **19**, confirm our hypothesis that a constrained, cyclic peptide can mimic the bioactive conformation more closely than a flexible linear peptide. The synthesis of RGD-containing peptides based on a structurally orientated design, led to a series of cyclic penta- and hexapeptide analogues demonstrating differentiated inhibitory profiles towards laminin P1-fragment and vitronectin mediated cell adhesion and fibrinogen binding to the isolated $\alpha_v\beta_3$ and $\alpha_{IIb}\beta_3$ receptor [45].

In the cell adhesion tests, it was found that ring contraction from cyclic

Table 2. Inhibitory capacity (IC_{50}) of cyclic penta- and hexapeptides towards cell adhesion to laminin P1-fragment [39]

Peptide[a]		IC$_{50}$ towards		
		HT 1080 (μM)	A 375 (μM)	HBL-100 (μM)
c(RGDFVA)	**1**	120	>120	95
c(RGDFVG)	**2**	120	>120	55
c(RGDFVA)	**3**	20	79	31
c(RGDFLA)	**4**	31	91	30
c(RGDFPG)	**5**	8	35	30
c(RGDFPA)	**6**	10	65	15
c(RGDFPG)	**7**	5	100	35
c(RGDFPA)	**8**	20	90	8
c(RGDFVA)	**9**	36	125	11
c(RGDFLA)	**10**	48	122	47
c(RGDFVG)	**11**	28	89	29
c(RGDFLG)	**12**	32	69	29
c(RGDFV)	**13**	18	114	25
c(RADFV)	**14**	>120	>120	>120
c(RGDFV)	**15**	49	>120	20
c(RGDFV)	**16**	0.2	1.0	0.1
c(RADFV)	**17**	18	>120	28
c(RGDFV)	**18**	1.0	1.9	0.9
RGDFV	**19**	92	29	42
GRGDS		5	8	4

[a]The amino acid abbreviations are given in the IUPAC one letter code. For D-amino acids bold letters are used. R = Arg, G = Gly, D = Asp, F = Phe, V = Val, A = Ala, S = Ser, P = Pro.

hexapeptides to pentapeptides, resulted in two highly active (**16, 18**) compounds and one selective peptide (**18**). Further investigations [44] showed that as far as the fibrinogen binding to the isolated receptors ($\alpha_v\beta_3$ and $\alpha_{IIb}\beta_3$) is concerned, the same pentapeptides **16** and **18** are selectively active to the vitronectin receptor [45].

Surprisingly, the conformational orientation of the RGD-sequence determined in active cyclic pentapeptides is also present within less active hexapeptides. Correlation between structure and activity indicated that ring size, but not necessarily the location of the RGD-sequence in a turn or elongated conformation, increases affinity for receptor binding.

We demonstrated that the cyclic pentapeptides **16** and **18** inhibit the laminin P1-fragment and vitronectin mediated adhesion to three different tumor cell lines as well as the in vitro binding of fibrinogen to the isolated $\alpha_v\beta_3$ receptor which is over-expressed on metastatic tumor cells.

These peptides could offer a possibility to interfere with the metastatic cascade and help to reduce metastasis in cancer patients.

142

Acknowledgements

We gratefully acknowledge the financial support of the Deutsche Forschungsgemeinschaft and the Fonds der Chemischen Industrie. The technical assistance of Dr G. Schnorrenberg (Boehringer Ingelheim KG) in providing facilities for peptide purification and Dr W. König (Hoechst AG, Frankfurt/M) for amino acid analysis is also acknowledged. We thank E. Merck, Darmstadt for financial support and especially B. Diefenbach for the realization of some biological assays. MG and GM gratefully acknowledge doctoral fellowships form the Studienstiftung des Deutschen Volks and the Fonds der Chemischen Industrie, respectively.

References

1. Liotta LA, Rao CN, Wewer UM. Ann Rev Biochem 1986;55:1037–1057.
2. Liotta LA. Caner Res 1986;46:1–7.
3. Hynes RA, Lander AD. Cell 1992;68:303–322.
4. Terranova VP, Hujanen ES, Martin GR. J Natl Cancer Inst 1986;77:311–316.
5. Springer TA. Nature 1990;346:425–434.
6. Albelda SM, Buck CA. FASEB J 1990;4:2868–2880.
7. Mould AP, Komoriya A, Yamada KM, Humphries MJ. J Biol Chem 1991;266:3579–3585.
8. Hynes RA. Cell 1987;48:549–554.
9. Ruoslathi E, Pierschbacher MD. Cell 1986;44:517–518.
10. D'Souza SE, Ginsberg MH, Plow EF. TIBS 1991;16:246–250.
11. Ruoslathi E, Pierschbacher MD. Science 1987;238:491–497.
12. Albelda SM, Motto SA, Elder DE, Stewart RM, Damjanovich L, Herlyn M, Buck CA. Cancer Res 1990;50:6757–6764.
13. Wayner EA, Carter WG. J Cell Biol 1987;105:1873–1884.
14. Cheresh DA, Smith JW, Cooper HM, Quaranta V. Cell 1989;57:59–69.
15. Terranova VP, Roa CN, Kalebic T, Morgulies M, Liotta LA. Proc Natl Acad Sci USA 1983;80:444–448.
16. Cheresh DA, Spiro RG. J Biol Chem 1987;36:17703–17711.
17. Humphries MJ, Yamada KM, Olden K. J Clin Invest 1988;81:782–790.
18. Gehlsen KR, Argraves WS, Pirschbacher MD, Ruoslathi E. J Cell Biol 1988;106:925–930.
19. Acharya R, Fry E, Stuart D, Fox G, Rowlands D, Brown F. Nature 1989;337:709–716.
20. Logan D, Abu-Ghazaleh R, Blakemore W, Curry S, Jackson T, King A, Lea S, Lewis R, Newman J, Parry N, Rowlands D, Stuart D, Fry E. Nature 1993;362:566–568.
21. Wistow G, Turnell B, Summers L, Slingsby C, Moss D, Miller L, Lindley P, Blundell T. J Mol Biol 1983;107:175–202.
22. Leahy DJ, Hendrickson WA, Aukhil I, Erickson HP. Science 1992;258:987–991.
23. Main LA, Harvey TS, Baron M, Boyd J, Campbell ID. Cell 1992;71:671–678.
24. Baron M, Main AL, Driscoll PC, Mardon HJ, Boyd J, Campbell ID. Biochemistry 1992;31:2068–2073.
25. Adler M, Carter P, Lazarus RA, Wagner G. Biochemistry 1993;32:282–289.
26. Adler M, Lazarus RA, Dennis MS, Wagner G. Science 1991;253:445–448.
27. Dalvit C, Widmer H, Bovermann G, Breckenridge R, Metternich R. Eur J Biochem 1991;202:315–321.
28. Cooke RM, Carter BG, Martin DMA, Murray-Rust P, Weir MP. Eur J Biochem 1991;202:323–328.
29. Saudek V, Atkinson RA, Lepage P, Pelton JT. Eur J Biochem 1991;202:329–338.
30. Saudek V, Atkinson RA, Pelton JT. Biochemistry 1991;30:7369–7372.

31. Chen Y, Pitzenberger SM, Garsky VM, Lumma PK, Sanyal G, Baum J. Biochemistry 1991;30: 11625–11636.
32. Senn H, Klaus W. J Mol Biol 1993;232:907–925.
33. Kessler H. Angew Chem 1982;94:504–520; Angew Chem Int Edn 1982;21:512–523.
34. Müller G, Gurrath M, Kessler H, Timpl R. Angew Chem 1992;104:341–343; Angew Chem Int Edn Engl 1992;31:326–328.
35. Kessler H, Haupt A, Will M. In: Purn TJ, Probst CL (eds) Computer-aided Drugdesign, Methods and Applications. New York: Marcel Dekker, 1989;461–484.
36. Havel TF. Biopolymers 1990;29:1565–1585.
37. Havel TF. Prog Biophys Mol Biol 1991;56:43–78.
38. Kessler H. In: Claasen V (ed) Trends in Drug Research. Amsterdam: Elsevier Science Publisher, 1990;73–84.
39. Gurrath M, Müller G, Kessler H, Aumailley M, Timpl R. Eur J Biochem 1992;210:911–921.
40. Pierschbacher MD, Ruoslathi E. J Biol Chem 1987;262:17925–17928.
41. Samanen J, Ali FE, Romoff T, Calvo R, Sørensen E, Bennett D, Berry D, Koster P, Vasko J, Powers D, Stadel J, Nichols A. In: Giralt E, Andreu D (eds) Peptides 1990: Proceedings of the 21st European Peptide Symposium. Leiden: ESCOM, 1991;781–783.
42. Nutt RF, Brady SF, Sisko JT, Ciccarone TM, Calton CD, Levy MR, Gould RJ, Zhang G, Friedman PA, Veber DF. In: Giralt E, Andreu D (eds) Peptides 1990: Proceedings of the 21st European Peptide Symposium. Leiden: ESCOM, 1991;784–786.
43. Aumailley M, Gurrath M, Müller G, Calvete J, Timpl R, Kessler H. FEBS Lett 1991;291:50–54.
44. Gurrath M. Thesis, TU München 1992.
45. The isolated receptor-tests have been performed by B. Diefenbach, E. Merck, Darmstadt.

The urokinase/urokinase receptor system — a new target for cancer therapy?

O. Wilhelm[1*], M. Schmitt[1], R. Senekowitsch[2], S. Höhl[1], S. Wilhelm[1], C. Will[1], P. Rettenberger[1], U. Reuning[1], U. Weidle[3], V. Magdolen[1] and H. Graeff[1]
[1]*Frauenklinik und* [2]*Institut für Nuklearmedizin der Technischen Universität München, München;*
[3]*Boehringer Mannheim, Penzberg, Germany*

Introduction

The serine protease urokinase-type plasminogen activator (uPA) in addition to tissue-type plasminogen activator (tPA) is one of the two enzymes known to specifically convert plasminogen to plasmin. This activation reaction plays a central role in extracellular proteolysis. The uPA molecule consists of three individual domains: 1) the N-terminal epidermal growth factor-like domain (EGF-like domain or GFD) with structural homology to the epidermal growth factor (EGF) [1]; 2) followed by the kringle domain whose secondary structure is related to the kringle domains of prothrombin [2], Hageman factor [3], tPA [4] and hepatocyte growth factor [5]; 3) the serine protease domain homologous to numerous other serine proteases [6]. The EGF-like domain contains the binding site (amino acids 17—32) which mediates the binding to the urokinase receptor (uPAR, also recently classified as CD 87) [7]. The kringle domain is known to interact with heparin and benzamidine [8,9]; the serine protease domain represents the catalytic domain of uPA. uPA generates the protease plasmin by cleaving the peptidyl bond $Arg^{560} - Val^{561}$ of plasminogen [10]. uPA is synthesized and secreted as an inactive proenzyme (pro-uPA, single-chain uPA) in a variety of normal and malignant cells. In solution or upon binding to the high affinity uPA receptor, uPAR, on the cell surface [11], pro-uPA may be cleaved by plasmin, kallikrein, cathepsin B/L or the nerve growth factor-γ to give rise to the active two-chain form of uPA (HMW-uPA) [12—15]. Pro-uPA may also be cleaved by thrombin at $Arg^{156} - Phe^{157}$, which results in the inactive two-chain form of uPA [16]. This inactivation of uPA by thrombin is accelerated 70-fold in the presence of thrombomodulin, a cell surface receptor for thrombin present on the endothelial cell surface [17].

uPAR is a heavily glycosylated unique protein (with an apparent molecular mass of about 50—60 kDa as estimated by sodium dodecyl sulphate-polyacrylamide gel

---

Address for correspondence: Dr Olaf Wilhelm, Frauenklinik der Technischen Universität München, Klinikum rechts der Isar, Ismaningerstraße 22, D-81675 München, Germany. Tel.: +49-89-41402493. Fax: +49-89-41805146.

electrophoresis (SDS-PAGE)) which is attached to the cell membrane via a glycosyl-phosphatidylinositol (GPI) anchor [18]. uPAR is composed of three internal repeats of approximately 92 amino acids which are each characterized by a distinct pattern of cysteine residues. A growing number of GPI-anchored proteins, e.g., the murine Ly-6 antigen, the squid brain glycoprotein Sqp-2 or the membrane inhibitor of reactive lysis, CD59, have also been identified to contain this characteristic cysteine motif. In contrast to uPAR, these proteins contain only one copy of this domain, indicating that uPAR may have arisen by triplication of an ancestral gene. The ligand (uPA) binding region is located in the N-terminal domain within the first 87 amino acids of uPAR [19]. Phospholipase C releases uPAR from the cell surface [20].

Receptor-bound HMW-uPA activates plasminogen equally well as free HMW-uPA into the broad spectrum serine protease plasmin, which also may be bound to a specific but low affinity receptor on the cell surface [21]. Plasmin then degrades extracellular matrix proteins like fibrin and fibronectin [22—24], activates procollagenases and generates HMW-uPA from pro-uPA. This cell-associated proteolytic activation system provides a potential cellular system for focal tissue degradation/remodeling and enables tumor cells to invade the surrounding matrix. Surprisingly, transgenic mice in which the uPA gene has been eliminated are viable, fertile and appear healthy [25] which implicates the nonessentiality of uPA expression in mammals under physiological conditions. Whether this applies to humans could not be shown so far. The importance of the uPA/uPAR system under pathophysiological conditions has been demonstrated by a number of studies.

Primary tumors of the breast, ovary, prostate, bladder and gastrointestinal tract contain high amounts of uPA and/or PAI-1 (plasminogen activator inhibitor 1), compared to benign control tissues. The elevated levels of PAI-1 in tumor tissues may be important for reimplantation of circulating tumor cells. Increased uPA and PAI-1 content in tumor tissue may serve as prognostic marker to predict relapse-free and/or overall survival in patients with cancer of the breast, ovary and the gastrointestinal tract [26,27].

In vitro studies with certain cancer cell lines have demonstrated that their invasive capacity depended on uPA [28]. It has also been shown that noninvasive, non-uPA expressing cells become invasive upon transformation with an expression plasmid encoding uPA [29]. These and other findings have underlined the important role of uPA and uPAR in the invasion/metastasis process of cancer cells. The question arises whether it is possible to efficiently interfere in vivo with this protease/receptor system and thus identify new therapeutic agents for treatment of human cancer.

The interaction of urokinase with the urokinase receptor

The receptor-binding region of the uPA molecule is located within the EGF-like domain, GFD (amino acid position 1—44) [7]. A peptide comprising residues 10—32 of uPA has been shown to bind to cell surface uPAR [7]. uPA molecules which lack the EGF-like domain, e.g., the low-molecular-weight uPA (LMW-uPA), are unable

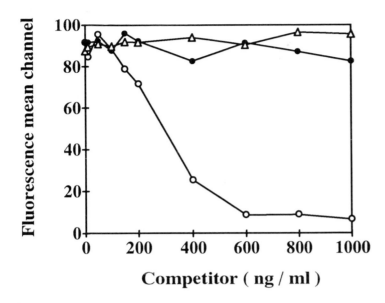

Fig. 1. Inhibition of binding of FITC-labeled pro-uPA to uPAR on PMA-stimulated U 937 cells by rec-uPAR$_{1-277}$ (open circles), the chymotrypsin-mediated fragment of rec-uPAR$_{1-277}$, representing amino acids 1-87, (open triangles) and the reduced and alkylated rec-uPAR$_{1-277}$ (closed circles). Experimental details are described in [43].

to interact with uPAR [28]. The ligand-binding domain in uPAR is located within the first, N-terminal located, repeat (amino acids 1–92) [19]. Limited digestion of uPAR with chymotrypsin generates a fragment (amino acids 1–87) which still retains the capability to bind to uPA, but is, however, not able to efficiently compete with uPA for binding to uPAR (Fig. 1). In contrast, addition of a 4-fold molar excess of a truncated, soluble uPAR variant (amino acids 1–277, rec-uPAR$_{1-277}$) missing the GPI-anchor produced in Chinese hamster ovary (CHO)-cells, completely blocks the binding of uPA to cell surface-associated uPAR (Fig. 1). The conformation of uPAR seems thus to be crucial for uPA binding since the reduced and alkylated form of rec-uPAR$_{1-277}$ does not block the binding of uPA to cell surface-bound uPAR (Fig. 1) [38].

The apparent molecular mass of uPAR varies from 50–65 kDa as judged by SDS-PAGE [30]. This molecular mass heterogeneity is due to N-linked glycosylation; the deglycosylated uPAR has an apparent molecular weight of 35 kDa only. Alteration of the glycosylation site at Asn^{52} to Gln within the uPA-binding domain by in vitro mutagenesis resulted in an active mutant devoid of carbohydrate in the ligand-binding domain. However, an about 5-fold decrease in affinity to uPA was observed when compared to wild type uPAR [31]. Treatment of cells expressing soluble uPAR with Tunicamycin or alteration of all five glycosylation sites of uPAR by site-directed mutagenesis did not result in uPA receptor molecules with ligand-binding capability [31].

uPA bound to its receptor is poorly, if at all, internalized by the cell. However, the ternary complex consisting of uPAR, uPA and its inhibitor, PAI-1, which is formed either by binding of a soluble uPA/PAI-1 complex to cell surface uPAR or by binding of PAI-1 to uPA already bound to uPAR [32], is readily internalized [38]. Recently it has been demonstrated that the multiligand α_2-macroglobulin receptor/low density lipoprotein-receptor-related protein (α_2MR/LRP) is involved in the binding and endocytosis of the ternary uPAR/uPA/PAI-1 complex [33]. Within the cell, uPA/PAI-1 separates from uPAR and is subsequently degraded while uPAR is most likely recycled, at least in part, to the cell surface [34]. A specific downregulation of uPAR was observed when cells were incubated with uPA/PAI-1 complexes [35]. The binding of uPA to uPAR induces a dose-dependent tyrosine phosphorylation of a 38 kDa protein which is associated with uPAR. The activation of a tyrosine kinase may reflect the initiation of a signal transduction pathway and indicates that uPA, besides its role as an extracellular protease, is implicated in intracellular signal transduction [36]. Further detailed studies of the internalization process and its role will be necessary to evaluate this possibility.

Expression of urokinase and urokinase receptor in cancer cells

uPA and uPAR are expressed in a variety of normal and malignant cells [25]. uPA can be quantified in cell extracts, cell supernatants, tumor tissues and biological fluids

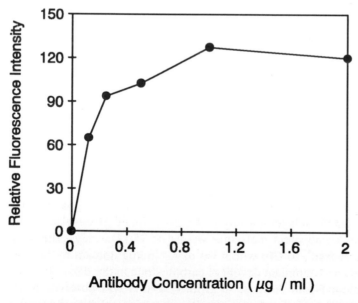

Fig. 2. Determination of cell-surface uPAR of OV-MZ-6 by flow cytofluorometry (FACScan, Becton Dickinson) by reaction with uPAR-directed monoclonal antibody # 3936 (American Diagnostica) followed by FITC-anti-mouse IgG. Relative fluorescence is expressed as mean fluorescence channels. Unspecific binding in the presence of a 60-fold molar excess of unlabeled pro-uPA was subtracted from each data point. Experimental details are described in [43].

by sensitive enzyme-linked immunosorbent assays (ELISAs) or by determination of uPA activity with specific synthetic chromogenic substrates. uPAR is assessed on normal and on tumor cells by several techniques. Initially, quantitative determination of uPAR on cells has been achieved using radioiodinated uPA [37]. A direct detection of uPAR on living tumor cells involves fluorescently labeled uPA and subsequent quantification by flow cytometry [38]. Laser-based flow cytofluorometry (FACScan) (Fig. 2) and confocal laser scanning microscopy (CLSM) (Fig. 4) using monoclonal antibodies directed to uPAR which in turn are quantified by fluorescently labeled second antibody, are also useful for detection of uPAR on living tumor cells [38].

To study possible new therapeutical uPA/uPAR reagents to be eventually administered to cancer patients, we characterized six different human ovarian cancer cell lines established from patients with cystadenocarcinoma of the ovary for expression of uPA, PAI-1 and uPAR. All six cancer cell lines synthesized uPA, PAI-1 and uPAR as studied by ELISAs, flow cytofluorometry and determination of the cell-associated uPA and plasmin activity. Immunohistochemically, uPA and uPAR were visualized in all six ovarian cancer cell lines by applying monoclonal antibodies directed to uPA and uPAR, respectively. As an example, the staining pattern for uPA and uPAR of the ovarian cancer cell line OV-MZ-11 is shown in Fig. 3. The localization of uPAR expressed in human ovarian cancer cells was also studied by CLSM (Fig. 4). uPAR is present on the cell surface and in the cytoplasm of OV-MZ-11.

All six cell lines displayed invasive capacity as tested in an in vitro invasion assay using Matrigel as the extracellular matrix. The two most highly invasive cell lines (OV-MZ-6, OVCAR-3) and, as a control, the very weak invasive cell line OV-MZ-11, were injected intraperitoneally into nude mice. The two invasive cell lines formed considerable tumor masses throughout the abdominal cavity, including metastases of the liver and the intestinal tract within a period of 4 weeks. OV-MZ-11 was not tumorigenic at all.

Inhibition of proliferation, invasion and tumor growth by uPA/uPAR derivatives

The expression of uPA by human ovarian carcinoma was first described in 1976 by Astedt and Holmberg [39]. Since then evidence has accumulated that the uPA/uPAR system is involved in cell proliferation, invasion and metastasis of tumor cells. In vitro, exogenous uPA stimulates clonal growth of HL-60 cells [40] and proliferation of the prostate cancer cell line CLL 20.2 [41]. In the latter case, the stimulatory effect of uPA was abolished after addition of the amino-terminal fragment (ATF) of uPA. In SaOS-2 cells, the fucosylated growth factor-like domain of uPA (amino acids 1—44 of uPA, GFD) elicited a mitogenic response. Surprisingly, the same domain, expressed in Escherichia coli (E. coli) without being glycosylated, exhibited no growth factor-like activity [42]. The addition of functionally active soluble recombinant uPAR (rec-uPAR$_{1-277}$) to human ovarian cancer cells inhibited uPA-dependent cell proliferation in a concentration-dependent manner by binding to uPA (Fig. 5). Recombinant soluble uPAR seems to function as a scavenger for uPA by

150

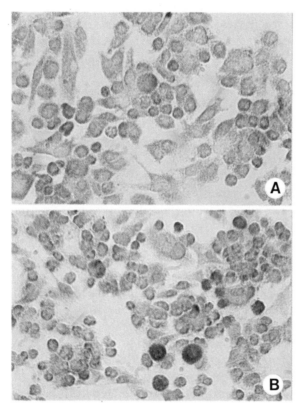

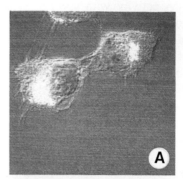

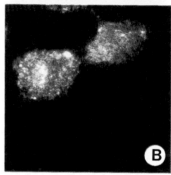

Fig. 3. Immunocytochemical detection of uPA and uPAR in the human ovarian cancer cell line OV-MZ-11. Cells were grown on chamber slides, fixed with paraformaldehyde and then stained with monoclonal antibody # 3689 (American Diagnostica) directed to uPA (A) and monoclonal antibody #3936 (American Diagnostica) against uPAR (B), respectively. Staining was visualized by the APAAP method as outlined in [57]. The nuclei were stained with hematoxylin (blue), the antibody-specific staining appears in red.

Fig. 4. Visualization of uPAR on cells of the human ovarian cancer cell line OV-MZ-6 by confocal laser scanning microscopy. Cells were grown on chamber slides, reacted with monoclonal antibody directed to uPAR (# 3936) and then fixed with paraformaldehyde. uPAR was visualized by reaction of bound #3936 with FITC-goat-anti-mouse IgG. 4A: Transmission light image (Nomarski optics). 4B: Fluorescence pattern (extended focus representing the superimposed fluorescence images of eight consecutively staged sections of the cells). Details of the experiments are described in [38]. White: high intensity, red: low intensity. Scale bar: 10 μm.

forming a uPA/uPAR complex and thus inhibiting the binding of endogenously produced uPA to cell surface uPAR and, in consequence, blocking the stimulatory effect of uPA on the tumor cells [43]. In an in vitro invasion assay, using Matrigel as the extracellular matrix, the addition of rec-uPAR$_{1-277}$ reduced the invasive capacity of human ovarian cancer cells up to 75% (Fig. 6) [43]. These results substantiate findings of others using monoclonal antibodies directed to uPA or competitive peptides comprising amino acids 10–32 of uPA that reagents directed to the uPA/uPAR system do in fact inhibit the in vitro invasion capacity of cancer cells [44,45].

To test whether rec-uPAR$_{1-277}$ is not only efficient in inhibiting tumor cell proliferation and invasion in vitro but also in vivo, the rec-uPAR$_{1-277}$ was injected intraperitoneally into nude mice once a week over a period of 5 weeks (200 µg rec-uPAR$_{1-277}$/week per mouse). The mice were then sacrificed, the tumor masses surgically removed and the tumor weight determined. As shown in Fig. 7, no apparent difference in tumor growth between treated and untreated mice was observed at the dosage used. Additional experiments with increased concentrations of rec-uPAR$_{1-277}$ may be necessary to evaluate a possible effect on locoregional tumor growth at a higher dosage of the recombinant protein. In this respect it has to be noted that in ascitic fluids of patients with cystadenocarcinoma of the ovary surprisingly high levels of a soluble form of the uPAR have been detected [46]. So far there is no explanation for this phenomenon of uPAR to be shed from cancer cells into the ascitic fluid.

The successful in vitro inhibition of invasion of human cancer cells has been demonstrated with monoclonal antibodies directed to uPA or peptides comprising amino acid residues 10-32 of uPA (summarized in Table 1). In another rather different experiment, the highly invasive prostate cancer cell line PC-3 was transformed with an expression plasmid which triggered PC-3 cells to secrete high levels of a proteolytically inactive uPA variant. After injection of this transformed cell line into nude mice it was observed that tumor cell metastasis was efficiently inhibited, however, without any influence on growth of the primary tumor [47].

Alternative strategies to inhibit invasion and tumor growth

A fairly new approach to inhibit cellular effects is the use of antisense mRNA oligonucleotides in order to prevent synthesis of certain proteins by the cell. We have only recently produced such antisense mRNA oligonucleotides directed to the mRNA of uPA [48]. These antisense oligonucleotides suppressing uPA expression significantly reduced tumor cell invasion of human ovarian cancer cells [48]. In another approach Ossowski et al. demonstrated that tumorigenic cancer cells showed a decrease in metastatic behavior after transformation with an antisense plasmid inhibiting urokinase receptor expression [56]. Another feasible approach to be tested at present is the repression of urokinase synthesis by agents blocking transcriptional and translational factors which are known to be involved in urokinase expression. Analysis of the directed effect of synthetic protease inhibitors on tumor growth, tumor cell invasion and proliferation through reduction of urokinase activity e.g., by inhibitors like 4-substituted benzo[b]thiophene-2-carboxamidines [49] or by inactivation of uPA via thrombin/thrombomodulin complexes [50], are still in their infancy and await verification.

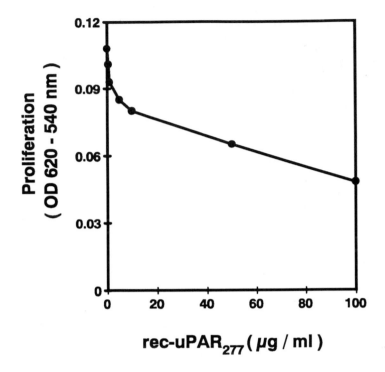

Fig. 5. Inhibition of proliferation of the human ovarian cancer cell line OV-MZ-6 by rec-uPAR$_{1-277}$. Rec-uPAR$_{1-277}$ was added in increasing concentration (0−100 μg/ml) to the cells and the proliferation rate was determined with a nonradioactive proliferation assay (Promega). Viability of the cells under the influence of rec-uPAR$_{1-277}$ was unaffected as determined by staining cells in parallel experiments with trypan blue [43].

Table 1. Literature review: Inhibition of proliferation, invasion and metastasis by blocking the uPA/uPAR interaction

Approach	Inhibitory effect on	Reference
Antibody against uPA	Metastasis	[44]
Antibodies against uPA	Metastasis	[51]
ATF	Proliferation	[52]
uPA peptides	Invasion	[53]
PAI-2	Invasion	[54]
Inactive uPA	Invasion	[55]
ATF, uPA peptides	Proliferation	[42]
Inactive uPA	Metastasis	[47]
uPA peptides	Invasion	[45]
Soluble uPAR	Invasion	[43]
uPAR antisense plasmid	Metastasis	[56]
uPA antisense oligonucleotides	Invasion	[48]

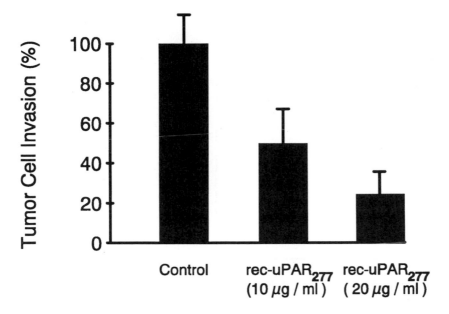

Fig. 6. Inhibition of invasion of human ovarian cancer cells (OV-MZ-6) in an in vitro invasion assay as detailed in [43]. Matrigel as the extracellular matrix was located between an upper polycarbonate filter and a lower nitrocellulose filter. 10^5 cells in 1 ml culture medium with or without rec-uPAR$_{1-277}$ were applied on the top of the filter sandwich and incubated for 16 h at 37°C. The whole set-up was fixed with glutaraldehyde for 72 h, then dissembled and the filter stained, either with hematoxylin (in case of the upper filter) or with Giemsa's solution (in case of the lower filter). The invasion rate was determined by calculating the ratio of the adherent cells on the upper filter and of cells on the lower and those who have invaded into the Matrigel [43].

Conclusion

Over the last decade substantial evidence has accumulated that the uPA/uPAR interaction is an important factor in tissue-associated proteolysis under physiological and pathophysiological conditions. Successful approaches to inhibit uPA/uPAR-dependent tumor cell invasion and cell proliferation by affecting the uPA/uPAR interaction are: 1) monoclonal antibodies directed to uPA; 2) uPA peptides and the ATF of uPA; 3) proteolytically inactive uPA; 4) recombinant soluble uPAR lacking the GPI-anchor; 5) inhibitor PAI-2; and 6) uPA and uPAR antisense systems. uPA-expression in mammals does not seem to be essential since transgenic mice in which the uPA gene has been eliminated are viable; this opens a new, broad therapeutic window. However, we have to consider the possibility that tumor cells — after the uPA/uPAR system has been successfully blocked — will develop back-up systems which might render the invasion and proliferation process independent of the uPA/uPAR system comparable to multidrug-resistance to chemotherapy. Nevertheless, the results obtained so far are sufficiently encouraging to continue a refined analysis

154

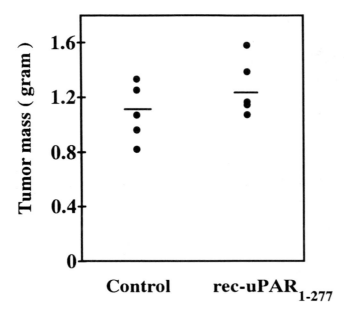

Fig. 7. Influence of rec-uPAR$_{1-277}$ on growth of human ovarian cancer cell line OV-MZ-6 in nude mice. 2×10^7 cancer cells/mouse (at the beginning of the experiment) and 200 μg rec-uPAR$_{1-277}$/mouse (once per week, altogether five times) were administered intraperitoneally. Control mice received cancer cells only. After 5 weeks mice were sacrificed, the tumors surgically removed and the weight of tumor masses determined. Each group consisted of five mice. Bars represent mean values.

of the urokinase/urokinase receptor system. In view of this, the final chapter for putative new agents directed to uPA/uPAR designed for treatment of human cancer by blocking the tumor cell associated protease system still has to be written.

Acknowledgements

This work was supported by the Deutsche Forschungsgemeinschaft (Clinical Research Group GR 280/4-1). The authors gratefully acknowledge the skillful technical assistance of E. Schueren and R. Schnelldorfer. We thank Dr W.A. Guenzler, Gruenenthal, Stolberg, FRG, for providing recombinant pro-uPA. We also thank Dr Richard Hart, American Diagnostica, Greenwich, CT, USA, for the generous support of our studies.

References

1. Banyai L, Varadi A, Patthy L. FEBS Lett 1983;163:37–41.
2. Magnusson S, Petersen TE, Sottrup-Jensen L, Claeys H. In: Reich E et al. (eds) Proteases and Biological Controls. Cold Spring Harbor, NY: Cold Spring Harbor Laboratory, 1975;123–149.

3. McMullen BA, Fujikama K. J Biol Chem 1985;260:5328–5341.
4. Guenzler WA, Steffens GJ, Otting F, Kim S-M, Frankus E, Flohe L. Hoppe-Seyler's Physiol Chem 1982;362:1151–1165.
5. Nakamura T, Nishizawa T, Hagiya M, Seki T, Shimonishi M, Sugimura A, Tashiro K, Shimizu S. Nature 1989;342:440–443.
6. Steffens GJ, Guenzler WA, Otting F, Frankus E, Flohe L. Hoppe Seyler's Physiol Chem 1982; 363:1043–1058.
7. Appella E, Robinson EA, Ullrich SJ, Stoppelli M, Corti A, Casani G, Blasi F. J Biol Chem 1987; 262:4437–4440.
8. Gurewich V, Pannell R, Louie S, Kelley P, Suddith RL, Greenlee R. J Clin Invest 1984;73: 1731–1739.
9. Winkler M, Blaber M, Bennett G, Holmes WE, Veher G. Biotechnology 1985;3:990–998.
10. Lijnen HR, Zamarron C, Blaber M, Winkler ME, Collen D. J Biol Chem 1986;261:1253–1258.
11. Blasi F. Fibrinolysis 1989;2:73–84.
12. Schmitt M, Goretzki L, Jänicke F, Calvete J, Eulitz M, Kobayashi H, Chucholowski N, Graeff H. Biomed Biochem Acta 1991;50:737–741.
13. Kobayashi H, Schmitt M, Goretzki L, Chucholowski N, Calvete J, Kramer M, Günzler WA, Jänicke F, Graeff H. J Biol Chem 1991;266:5147-5152.
14. Goretzki L, Schmitt M, Mann KH, Calvete J, Chucholowski N, Kramer M, Günzler WA, Jänicke F, Graeff H. FEBS Lett 1992;297:112-118.
15. Wolf B, Vasudevan J, Henkin J, Gonias S. J Biol Chem 1993;268:16327–16331.
16. Ichinose A, Fujikama K, Suyama T. J Biol Chem 1986;261:3486–3489.
17. de Munk GAW, Groeneveld E, Rijken DC. J Clin Invest 1991;88:1680–1684.
18. Roldan AL, Cubellis MV, Masucci MT, Behrendt N, Lund LR, Danø K, Appella E, Blasi F. EMBO J 1990;9:467-474.
19. Behrendt N, Ploug M, Patthy L, Houen G, Blasi F, Danø K. J Biol Chem 1991;266:7842–7847.
20. Moller L, Ploug M, Blasi F. Eur J Biochem 1992;208:493–500.
21. Plow EF, Freaney DE, Plescia J, Miles LA. J Biol Chem 1986;103:2411–2420.
22. Danø K, Andreasen PA, Grondahl-Hansen J, Kristensen P, Nielsen LS, Skriver L. Adv Cancer Res 1985;44:139-266.
23. Wilhelm O, Hafter R, Henschen A, Schmitt M, Graeff H. Blood 1990;75:1673-1678.
24. Wilhelm O, Hafter R, Coppenrath E, Pflanz MA, Schmitt M, Babic R, Linke R, Gössner W, Graeff H. Cancer Res 1988;48:3507-3514.
25. Carmielet P, Schoonjans L, Kieckens J, Stassen JM, Collen D, Mulligan R. In: Molecular & Cellular Biology of Plasminogen Activation Meeting. Cold Spring Harbor, NY: Cold Spring Harbor Laboratory, 1993.
26. Jänicke F, Schmitt M, Ulm K, Gössner W, Graeff H. Lancet 1989;2:1049.
27. Schmitt M, Jänicke F, Graeff H. Fibrinolysis 1992;6(suppl 4):3–26
28. Duffy MJ. Fibrinolysis 1993;7:295–302.
29. Cajot J-F, Schleuning W-D, Medcalf R, Bamat J, Testuz J, Liebermann L, Sordat B. J Cell Biol 1989;109:915–925.
30. Blasi F, Stoppelli P, Cubellis M. J Cell Biochem 1986;32:179–86.
31. Moller L, Pollaenen J, Ronne E, Pedersen N, Blasi F. J Biol Chem 1993;268:11152–11159.
32. Ellis V, Wun TC, Behrendt N, Ronne E, Danø K. J Biol Chem 1990;265:9904–9908.
33. Nykjaer A, Petersen C, Moller B, Jensen P, Moestrup S, Holtet Th, Etzerodt M, Thogersen H, Munch M, Andreasen P, Gliemann J. J Biol Chem 1992;267:14543–14546.
34. Cubellis MV, Wun TC, Blasi F. EMBO J 1990;9:1079-1085.
35. Olson D, Pollanen J, Hoyer-Hansen G, Ronne E, Sakaguchi K, Wun T, Appella E, Danø K, Blasi F. J Biol Chem 1992;267:9129–9133.
36. Dumler I, Petri T, Schlening WD. FEBS Lett 1993;322:37-40.
37. Stoppelli M, Corti A, Soffientini A, Cassani G, Blasi F, Assoian R. Proc Natl Acad Sci USA 1985;82:4939–4943.

156

38. Chucholowski N, Schmitt M, Rettenberger P, Schüren E, Moniwa N, Goretzki L, Wilhelm O, Weidle U, Jänicke F, Graeff H. Fibrinolysis 1992;6(suppl 4):95-102.
39. Astedt B, Holmberg L. Nature 1976;262:595-597.
40. Berdel WE, Wilhelm O, Schmitt M, Maurer J, Reufi B, von Marschall Z, Oberberg D, Graeff H, Thiel E. Int J Oncol 1993;3:607-613.
41. Kirchheimer J, Christ G, Binder B. Eur J Biochem 1989;181:103-107.
42. Rabbani S, Mazar A, Bernier S, Haq M, Bolivar I, Henkin J, Goltzman D. J Biol Chem 1992;267: 14151−14156.
43. Wilhelm O, Weidle U, Höhl S, Rettenberger P, Schmitt M, Graeff H. FEBS Lett 1994;337:131−134.
44. Ossowski L, Reich E. Cell 1983;35:611-619.
45. Kobayashi H, Ohi H, Shinohara H, Sugimura M, Fujii T, Terao T, Schmitt M, Goretzki L, Chucholowski N, Jänicke F, Graeff H. Br J Cancer 1993;67:537-544
46. Pedersen N, Schmitt M, Hoyer-Hansen G, Ronne E, Danø K, Kuhn W, Jänicke F, Blasi F. J Clin Invest 1993;92:2160−2167.
47. Crowley CW, Cohen RL, Lucas BK, Liu G, Shuman MA, Levinson AD. Proc Natl Acad Sci USA 1993;90:5021-5025.
48. Wilhelm O, Weidle U, Will C, Höhl S, Rettenberger P, Brünner N, Senekowitch R, Schmitt M, Graeff H. In: Molecular & Cellular Biology of Plasminogen Activation Meeting. Cold Spring Harbor, NY: Cold Spring Harbor Laboratory, 1993.
49. Towle M, Lee A, Maduakor E, Schwartz E, Bridges A, Littlefield B. Cancer Res 1993;53: 2553−2559.
50. Wilhelm S, Wilhelm O, Schmitt M, Möbus V, Kreienberg R, Graeff H. Gyn Rundschau 1993;33 (suppl 1):202−203.
51. Hearing V, Law L, Corti A, Appella E, Blasi F. Cancer Res 1988;48:1270−1278.
52. Kirchheimer JC, Wojta J, Christ G, Binder BR. Proc Natl Acad Sci USA 1989;86:5424-5428.
53. Schlechte W, Murano G, Boyd D. Cancer Res 1989;49:6064−6069.
54. Baker M, Bleakley P, Woodrow G, Doe W. Cancer Res 1990;50:4676−4684.
55. Cohen R, Xi X-P, Crowley C, Lucas B, Levinson A, Shuman M. Blood 1991;78:479−487.
56. Ossowski L, Zelent A, Kook Y-H. Molecular & Cellular Biology of Plasminogen Activation Meeting. Cold Spring Harbor, NY: Cold Spring Harbor Laboratory, 1993

Prospects in diagnosis and treatment of breast cancer
M. Schmitt et al., editors

Reduction of metastatic tumor cells in bone marrow of breast cancer patients treated with monoclonal antibody

Günter Schlimok[1*], Klaus Pantel[2] and Gert Riethmüller[2]

[1]*Medical Clinic II, Zentralklinikum, 86156 Augsburg; [2]Institute for Immunology, University of Munich, 80336 München, Germany*

Abstract. The initial promise of monoclonal antibodies as major therapeutic agents in human epithelial cancer has not been realized. Inaccessibility of cells in solid tumors due to factors such as the nature of the vascular endothelia and high pressure in the tumor are primarily responsible for the failure of antibody therapy. Although new strategies employing recombinant antibodies and immunoglobulins designed to actively engage the immune system may prove beneficial, micrometastatic tumor cells (at the stage of minimal residual disease) are likely to be the only suitable targets for antibody therapy. The diagnostic approaches to identify these cells and their use for monitoring adjuvant immunotherapy is discussed.

For almost a decade murine monoclonal antibodies have been tried for the therapy of various types of advanced human solid tumors [1]. While the failure of this experimental therapy — a few anecdotal and transient remissions excepted — has become undeniable, explanation for it has remained contentious. Although it is possible that unmodified murine antibodies lack the capacity to kill epithelial tumor cells in vivo, it may simply be that in contrast to leukemic cells, malignant epithelial cells encased in solid tumor parenchyma are largely inaccessible for intravenously administered antibodies. Indeed, there is increasing evidence that penetration of macromolecular antibodies into solid tumor tissue is severely hampered by numerous obstacles. The multifold barriers of the tumor's vasculature and the peritumorous basement membrane itself, the dense mesh of interstitial proteoglycans as well as intracellular tight junctions of epithelial cells, and last but not least the high oncotic pressure within the tumor interstitium, leading to a convection of interstitial fluid towards the tumor periphery, all prevent extravasation and free diffusion of injected antibodies [2]. Furthermore, effector mechanisms like complement or cytotoxic effector cells while abundant in the stromal tissue are absent or decreased in the epithelial parenchyma itself. Similar accessibility problems will also be encountered with other macromolecular therapeutic agents, such as immunotoxins and bispecific antibodies.

In contrast, individual epithelial tumor cells dispersed in mesenchymal compartments during the early phase of metastasis appear more appropriate targets for i.v.

Address for correspondence: Dr G. Schlimok, II. Med. Klinik, Zentralklinikum, Stenglinstraße, 86156 Augsburg, Germany.

injected antibodies. Since these dispersed cells can now be identified immunocyto-chemically in bone marrow by monoclonal antibodies against epithelial cytokeratins (CK) [3,4], the attempt was made to monitor their eradication by analyzing sequential bone marrow aspirates from breast cancer patients during treatment with a cytotoxic murine monoclonal antibody. Further justification for the undertaking of this trial was derived from the finding that in breast cancer the presence of micrometastatic cells in bone marrow can now be taken as a prognostic indicator of a later clinical relapse [5–8].

The IgG3 antibody (ABL364) used in this study recognizes the Lewis Y antigen which is expressed on 60–80% of human breast carcinoma. It exerts not only a remarkably high complement dependent cytotoxicity of up to 90% against various human tumor cell lines in vitro but, when intravenously administered to patients, also attains high levels of cytotoxicity in serum as shown by ex vivo experiments with fresh, not inactivated serum taken from antibody-treated patients [9–11].

Eight patients with breast cancer presenting high numbers of epithelial cytokeratin-positive cells in bone marrow ($>20/4 \times 10^5$ nucleated marrow cells) were treated in a randomized fashion during 2 weeks with either 6×100 mg of ABL364 or with human serum albumin (HSA) as a placebo.

The monitoring data are shown in Table 1. Three aspirations were performed at the following time points: the first prior to commencement of the antibody infusion, the second on day 15 and the third on day 60. Of the six patients treated with 6×100 mg antibody four showed a distinct decrease or disappearance of cytokeratin-positive Lewis Y-positive cells. Interestingly, the cytokeratin-positive cells observed 60 days after treatment appeared to be negative for the Lewis Y-antigen, suggesting a selection process favoring clonal expansion of cells lacking the target antigen. The

Table 1. Monitoring cytokeratin-positive tumor cells in bone marrow of breast cancer patients under treatment with monoclonal antibody ABL364

Patient	Treatment on:		No. of CK-positive cells per 4×10^5 bone marrow cells on indicated days			
	Day 0	Day 60	Day $\leq$0	Day 15	Day 60	Day 75[c]
	PLACEBO	ABL364				
K.U.[a]	+	+	320	300	380	22
H.H.[a]	+	+	1051[b]	900[b]	259[b]	924[b]
	ABL364					
B.P.	+		860	28[b]	47[b]	
D.G.	+		27	0	0	
R.O.	+		22	10	0	
W.H.	+		114	2[b]	12[b]	
M.W.	+		349[b]	427[b]	n.d.	
S.H.	+		62[b]	353[b]	950[b]	

[a]Crossing-over to Mab on day 60; [b]All cells were LEWIS Y antigen negative in parallel staining; [c]On day 75 an analysis was carried out on crossover patients only, i.e., on day 15 after Mab infusion.

remaining two patients presenting cytokeratin-positive Lewis Y-negative metastatic cells did not respond with a noticeable decrease of these cells. Similarly, no decrease or eradication of cells was observed in patients receiving human serum albumin only, regardless of antigenicity.

In the placebo patients a crossing-over to the treatment arm was performed as permitted by the protocol. While the first patient (K.U.) exhibited a rather stable number of cytokeratin-positive Lewis Y-positive cells over a period of 60 days, a drastic decrease of >95 percent was found on day 75, about 48 h after the last infusion of antibody. The second placebo patient also crossed over to the treatment arm, presented initially with >1000 metastatic epithelial cells lacking the Lewis Y antigen altogether. Here, after antibody treatment at day 75, no decrease of tumor cells could be found.

Except for some nausea, no major adverse effects were observed. In particular, no anaphylactic reactions were noted during the 2-week period of treatment. Only low titers of human antimouse antibodies were found in all treated patients.

As to the six patients with clinical manifest metastases, no objective regression of visible metastatic lesion could be ascertained by conventional diagnostic techniques such as x-ray or CT analysis. In the two patients with no evidence of clinical metastasis (Mo) no relapse of the disease has occurred so far.

The presented work demonstrates that the administration of monoclonal antibody ABL364 leads to a reduction or eradication of micrometastatic cells from bone marrow. It is suggested that monitoring of micrometastatic cells might be used as a novel surrogate measure of adjuvant therapies in minimal residual cancer.

To analyze the clinical efficacy of monoclonal antibodies against micrometastatic cells, a prospective randomized adjuvant trial was initiated in colorectal cancer. Only patients in stage Dukes C who had undergone curative surgery and were free of macroscopic disease were admitted to this study.

The intravenously injected monoclonal antibody was directed against the 17-1A antigen, a 37 kDa glycoprotein of the cell membrane of normal and malignant epithelial cells. This monoclonal antibody induced antibody-dependent cellular cytotoxicity (ADCC) with human effector cells and prevented outgrowth of xenotransplanted human tumor cells in athymic mice [12,13].

One hundred and eighty-nine patients with colorectal cancer of stage Dukes C were randomly assigned to either an observation regimen or to postoperative treatment with 500 mg antibody 17-1A, followed by 4 monthly infusions of 100 mg of the same antibody. A balance of risk factors in the two groups was achieved by a dynamic randomization procedure [14].

After a median follow up of 5 years, antibody treatment reduced the overall death rate by 30% (Cox proportional hazard: $p = 0.04$, log-rank: $p = 0.05$) and decreased the recurrence rate by 27% (Cox proportional hazard: $p = 0.03$, log rank: $p = 0.05$). The effect of antibody was most pronounced on the manifestation rate of distant metastases as the first event (Cox proportional hazard: $p = 0.0014$, log rank: $p = 0.002$), an effect which was not seen for local relapses (Cox proportional hazard: $p = 0.74$, log rank: $p = 0.67$). Toxic effects of monoclonal antibody 17-1A were

160

infrequent and only minor, consisting mainly of mild constitutional and gastrointestinal symptoms. During 371 infusions four anaphylactic reactions were observed, all controllable by intravenous steroids and none necessitating hospitalization.

In conclusion, adjuvant treatment with 17-1A antibody extends life and prolongs remission in patients with colorectal cancer of stage Dukes C. This study shows that the principle of passive antibody therapy is valid when applied to minimal residual colorectal cancer. Further prospective randomized adjuvant trials are warranted to answer the question whether antibody treatment will also prolong survival in breast cancer patients.

Acknowledgements

This work was supported by Deutsche Krebshilfe, Bonn and Wilhelm Sander Stiftung, Neuburg, Donau.

References

1. Riethmüller G, Johnson JP. Curr Opinion Immunol 1992;4:647–655.
2. Jain RK. Cancer Res 1990;50:2747–2751.
3. Schlimok G, Funke I, Holzmann B et al. Proc Natl Acad Sci USA 1987;84:8672–8676.
4. Cote RJ, Rosen PP, Hakes TH, Sedira M et al. Am J Surg Pathol 1988;12:333–340.
5. Cote RJ, Rosen PP, Lesser ML et al. J Clin Oncol 1991;9:1749–1756.
6. Mansi, JL, Berger U, Easton D et al. Br Med J 1987;295:1093–1096.
7. Schlimok G, Lindemann F, Holzmann K et al. Proc Am Soc Clin Oncol 1992;11:102.
8. Diel IJ, Kaufmann M, Goerner R et al. J Clin Oncol 1992;10:1534–1539.
9. Steplewski Z, Blaszezyk Thurin M et al. Hybridoma 1990;9:201–210.
10. Scholz D, Lubeck M, Loibner H et al. Cancer Immunol Immunother 1991;33:153–157.
11. Schlimok G, Riethmüller G, Pantel K et al. Proc Am Soc Clin Oncol 1991;10:212.
12. Herlyn M, Steplewski Z, Herlyn D, Koprowski H. Proc Natl Acad Sci USA 1979;76:1438–1442.
13. Herlyn DM, Steplewski Z, Herlyn MF, Koprowski H. Cancer Res 1980;40:717–721.
14. Riethmüller G, Schneider-Gädicke E, Schlimok G. et al. Ann Hematol 1993;67(suppl):102.

©1994 Elsevier Science B.V. All rights reserved
Prospects in diagnosis and treatment of breast cancer
M. Schmitt et al., editors

Intracellular and transmembranic tumor characteristics as targets for diagnosis and therapy: application of 16α-[123I]iodo-17β-estradiol and Mab 425 (EGF-R antibody)

M.W. Beckmann[1,5*], A. Scharl[2,5], H.G. Schnürch[1], K. Scheidhauer[3], R.P. Baum[4], D. Niederacher[1], J.R. Holt[5] and H.G. Bender[1]

[1]*Department of Obstetrics & Gynecology, Heinrich-Heine-Universität, Düsseldorf, Germany; [2]Department of Obstetrics & Gynecology, Universität zu Köln, Cologne, Germany; [3]Department of Nuclear Medicine, Universität zu Köln, Cologne, Germany; [4]Institute of Nuclear Medicine, Johann Wolfgang Goethe-Universität, Frankfurt a. M, Germany; [5]Department of Obstetrics & Gynecology, University of Chicago, Chicago, USA*

Introduction

The decision for hormone-, radiation- or chemo-therapy of patients with gynecological or breast cancer, is mainly based on stage and histopathological criteria [1—4]. In addition, functional tumor characteristics including steroid receptor status (estrogen (ER) and progesterone receptor (PgR)) proved to have prognostic significance and relevance for therapeutic decisions [2—5]. Large series of functional tumor characteristics, including factors involved in tumor proliferation, invasion and angioneogenesis, are currently under investigation [3,6—8]. These factors include membrane receptors for hormones and growth factors (LHRH-R, PRL-R, IGF-1-R, EGF-R, TGF-β-R, SS-R), enzymes, proteins, and other cytoplasmic factors (cathepsin D and B, urokinase plasminogen activator, plasminogen activator inhibitor, pS2, heat shock proteins, EGF, TGF-α, TGF-β, IGF-1) and Ki 67, CD31, p53, int-2, *ras*, and c-*myc* [8].

The increased understanding of tumor biology and advanced biotechnological tools foster novel approaches to cancer diagnosis and therapy. In this report, we present two different possibilities of site-directed diagnosis and/or therapy using two distinct biological characteristics of tumor cells:
— a radiolabeled ligand, 16α-[123I]iodo-17β-estradiol ([123I]E), directed to the intracellular estrogen receptor (ER) may be useful for tumor imaging and may also bear potential for therapeutic ER-mediated radiocytotoxicity;
— a monoclonal antibody, Mab 425, directed against the transmembranous epidermal growth factor receptor (EGF-R) for diagnostic scintigraphic tumor detection ([99Tc]Mab 425) and immunologic interaction for therapeutic use (Mab 425).

Address for correspondence: Matthias W. Beckmann, MD, Universitäts-Frauenklinik, Heinrich-Heine-Universität, D-40225 Düsseldorf, Germany. Tel.: +49-211-311-7500. Fax: +49-211-76-5425.

16α-[^{123}I]iodo-17β-estradiol ([^{123}I]E)

Strategies for diagnosis and therapy in which sex steroid receptor ligands serve as carriers for radionuclides are attractive because a high incidence of carcinomas of the female genital tract and the breast have an abundant expression of one or more of the receptor proteins [5,9]. Recent reports demonstrated the effectiveness of different radiolabeled estrogen receptor (ER) ligands, i.e., 16α-[^{18}F]fluoro-17β-estradiol ([^{18}F]E) [10] and [^{123}I]E [11,12] for imaging ER-rich breast tumors and their metastases in women. Furthermore, the in vivo imaging of ER-rich tumors bears the potential for site-directed radiotherapy of ER rich tumors including ovarian, endometrial or breast cancers [13,14]. This report focuses on the radiohalogenated estrogen receptor (ER) ligand, [^{123}I]E.

A

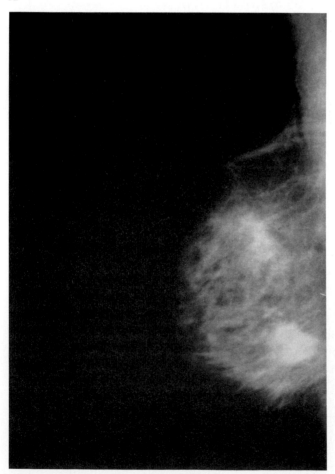

Fig. 1. A 58-year-old women with primary breast tumor in the right breast. A) mammography (medio-lateral): suspicious mass caudal.

B

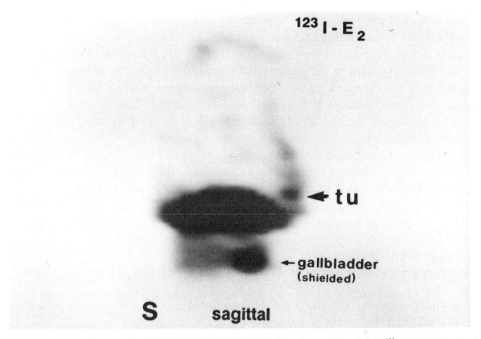

Fig. 1. (*continued*). B) sagittal SPECT-slice through the right breast 4-h p.i. of [^{123}I]E, showing the hot spot in the breast (arrow). High linear uptake and "hot" gallbladder already shield with lead [11].

Clinical trial with ER imaging in vivo

In a clinical trial [^{123}I]E was used as a receptor-specific radiopharmacon for scintigraphic tumor detection in 62 patients suspected of breast cancer [11,12]. The study was performed as a multicenter trial (five university hospitals) to validate the method and to overcome methodogical problems. Inclusion criteria were: suspicion of primary tumor, local recurrence or known metastatic disease. [^{123}I]E was produced by Amersham-Buchler, Braunschweig Germany. The applied dose was 111 MBq (3 mCi) diluted in 6–10 ml solvent containing less than 20 ng of hormone. The thyroid was not blocked with cold iodine and could therefore be used as in vivo quality control for the radiolabeling of estradiol. Acquisition of data was obtained by real time imaging of planar pictures, thoracic SPECT or whole body imaging (20 min to 24-h p.i., Fig. 1). A fast tracer elimination from the blood pool into the liver was seen, followed by biliary excretion allowing early imaging of the thorax due to low background activity, but resulting in difficult imaging conditions for the abdomen. Thirteen patients had to be excluded due to technical difficulties, lack of postoperative ER analyses or synchronous hormone therapy. In 42 patients (30 carcinomas, 12 benign lesions) the overall sensitivity was 66% (biochemically determined ER status cut-off: 10 fmol/mg soluble cytosolic protein). Some patients with breast cancer showed focal or diffuse uptake in the area of primary lymph drainage (parasternal, axillary) without any clinical tumor correlation demanding follow-up investigations. There was only one false-positive result in a receptor-negative carcinoma; thus, the noninvasive determination of the ER status seems to be feasible. The sensitivity of [^{123}I]E in the detection of primary breast cancer or metastases and recurrences is low compared to mammography and other methods; therefore, [^{123}I]E-scintigraphy cannot be used as a routine screening method.

Differentiation of malignant and benign tissue is difficult as both may have a positive ER status, as seen in fibricystic disease. Nevertheless, [^{123}I]E scintigraphy is a possible in vivo imaging technique for the detection of breast cancer depending on the ER status, and provides information about tumor localization. It may become a specific method for the noninvasive diagnosis of the ER status and may be helpful in follow-up studies. As a receptor-specific agent [^{123}I]E may give answers to questions of tumor heterogeneity and changes of the ER status during therapy.

Pharmacodynamic and pharmacokinetic of [^{123}I]E in different animal models

The information about specific local enrichment of the radiopharmacon bears the potential of [^{123}I]E for site-directed therapy ("molecular surgery"). [^{123}I]E retains its high affinity and specificity for the estrogen receptor and can be produced in quantities that are necessary for this purpose. Especially for therapeutic application, dosimetry calculations for target and nontarget organs require knowledge of the distribution over time of radioactivity in the body with regard to different routes of administration (i.v., i.a., i.p.). Pharmacodynamic and pharmacokinetic studies were performed with the sister nuclide derivate ^{125}I, because of its prolonged, and therefore better-to-measure half-life compared to ^{123}I [15—18] (Table 1).

The method of choice for these studies is the indicator dilution technique that allows monitoring of blood-tissue exchange of radioactivity in a continuous manner in the swine model [18,19]. The domestic swine is a pharmacokinetic model that closely resembles humans [19,20].

We have recently demonstrated that iodination of estradiol does not alter the excellent diffusibility of estradiol [18]. When iodoestradiol is injected into the artery of a nontarget organ, the majority of radioactivity exits the blood during first pass.

Table 1. Specific activity of different iodine nuclides

	$t_{1/2}$[d]	Maximum specific activity	
		[Ci/mol]	[Ci/mg]
^{123}I	0.5	230.000	8.900
^{125}I	60	2.200	17.3
^{131}I	8	16.100	123

Subsequently, with the change in concentration gradients, radioactivity is released back from the organ tissue into the venous blood. If this phenomenon, which has been observed in pig leg, also occurs in other organs, it has important clinical consequences: it will have influence on pharmacodynamics and radiation in nontarget organs. Furthermore, it is possible that iodoestradiol undergoes metabolism in different organs. Radioestradiol is rapidly metabolized by the liver during first pass as seen with the indicator dilution method [16]. Ninety-nine percent of [^{125}I]E is cleared after injection into the portal vein; water-soluble metabolites are then partly released into the blood and partly excreted into bile. After injection of [^{125}I]E into the external jugular vein, one third of the radioactivity is excreted in bile and two-thirds in the urine. More than 90% of the radioactivity in urine and bile is that of [^{125}I]E-glucuronide or [^{125}I]E-sulfate; only a very small fraction of the excreted radioactivity is from free ^{125}I. Furthermore, enterohepatic recirculation was examined. Radioactivity in bile collected from one swine after i.v. injection of [^{125}I]E and then infused into the proximal duodenum of a second swine is almost totally absorbed during passage through the intestine at 5—7 h after infusion. The reabsorbed radioactivity is cleared in the urine [16].

Natural estrogens are metabolized in the liver and in many other organs, as is fat, intestine and skin [21,22]. Effective metabolism of iodoestradiol in several peripheric organs and tissues into substances with lower or no affinity for ER or de-iodination, implies that during first pass of the radioestrogen through the circulation most of the pharmacon is inactivated. Consequently only a small part of the dose applied would be available for binding to ER. If no deactivation of the radioestrogen occurs, nontarget organs could act as depots that result in a prolonged inflow of the radiopharmacon into target tissue with the possible consequence of a cumulative and enhanced saturation of ER. Intravenous application of pharmaca is most important for clinical purposes. Before the pharmacon enters the body circulation, it has to pass the lung. In case of rapid clearance of halo-steroids from the blood into the tissue, a large portion of the injected activity would escape into the lung tissue during first pass through the capillary system. These problems should be taken into account. Up-to-date data about the single pass clearance of radioestradiol in various parenchymatous organs after i.v. or i.a. injection are not available, but are under current investigation [Scharl, Habilitation Köln 1992]. The results should be considered in human subjects when radiolabeled steroid receptor ligands are to be used in humans.

From another animal model, data are available about the i.p. application of [^{125}I]E [17,23]. Radioestrogens have the potential as adjunct agents against ovarian carcinomas using the intraperitoneal administration as a possible approach. Absorption from the peritoneal cavity of [^{125}I]E after i.p. application in rats with and without ovarian tumors and ascites were investigated and compared with the distribution of the radioactivity after i.v. injection. In the absence of ascites, 70% of the intraperitoneal dose was cleared into the intestine within 2 h after injection, indicating absorption from the peritoneal cavity. In the presence of ascites, clearance of intraperitoneal radioiodoestradiol was considerably slower; at 2 h after injection, 50% of the injected dose remained in the ascites, mostly as radioiodoestradiol. Uptake of radioactivity in estrogen receptor-rich tissues, e.g., uterus, after intraperitoneal injection was high (ratio about 20:1 over blood), regardless of the presence of ascites, but moderately lower than that observed after intravenous injection of radioiodoestradiol.

It is important to realize that the situation in the animal models (swine, mice or rat) do not necessarily mirror the situation in humans. In earlier experiments [16] we were able to demonstrate that the renally excreted metabolites of iodoestradiol in swine and man showed similarities, but had quantitative differences.

Cytotoxicity of [^{125}I]E in vitro

The success of in vivo imaging [10−12] highlights the potential for the use of such radiopharmaceutics especially iodoestradiol derivates ([I]E) for therapy. Additional rationale for the use of radiohalogenated ligands in therapy comes from the observation that the tagged-ligands cause selective cytotoxicity to ER-rich cells in vitro, when [^{125}I]E is used as the radiolabel [24,25]. However, because of their potential significance for cancer therapy, the mechanisms of the cytotoxicity from [^{125}I]-labeled ER ligands constitute an area of interest [26,27]. The cytotoxic potential of ^{125}I, when attached to deoxyuridine (UdR) and incorporated into the DNA backbone, has been shown to be much higher than that of [^{131}I]-UdR [28]. ^{125}I is an electron (Auger electron) emitter with very low energy, whereas ^{131}I does not produce Auger electrons, but decays by gamma emission [29]. The cytocidal effect of ^{125}I, which is much like the effect of high linear energy transfer (LET), is believed to be caused by the Auger electron shower from orbital electron capture [30]. Similar results are reported for other DNA-bound Auger electron-emitters, e.g., ^{77}Br [31] and ^{80}mBr [32]. For ^{125}I the cytotoxicity attributable to the emission of Auger electrons is believed to be related to DNA strand breaks that occur within 15 to 20 Å of the site of the radiodecay. Recently Woo et al. [27] reported that an association of a [^{125}I]-labeled monoclonal antibody ([^{125}I]17-1a) to the nuclear DNA of the human cancer cell line SW1116 was necessary to achieve specific target cell toxicity and was accompanied by dose-dependent chromosomal aberrations. In a standard colony-forming assay in vitro cytotoxicity of [^{125}I]E caused by the formation of [^{125}I]E − ER complex in the cell nucleus could be demonstrated [33,34]. The addition of [^{125}I]E to breast cancer cell cultures decreased the survival rate of ER-positive MCF-7 cells in a dose-dependent manner (Fig. 2). The decreased survival rate was prevented by the addition of competing excess molar radioinert ER ligand (diethylstilbestrol, DES); [^{125}I]E did not reduce survival in ER-negative MCF-7 cells. The [^{125}I]E-induced and ER-mediated cytotoxicity was accompanied by aberrations in the DNA components of the nuclei of the cells. These included chromatide- and chromosome-breaks, gaps, and triradial chromosome formation [33,34].

The radiocytotoxicity observed is compatible with receptor-mediated therapy because ER-rich cancers contain from 500 to 20,000 free receptor molecules per cell. Whereas current endocrine therapies are basically cytostatic, receptor-directed therapy

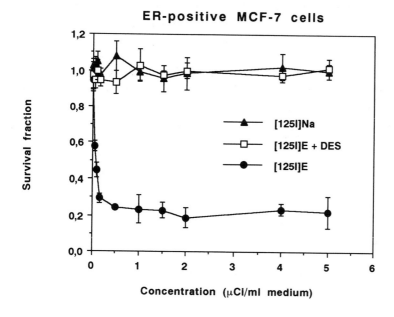

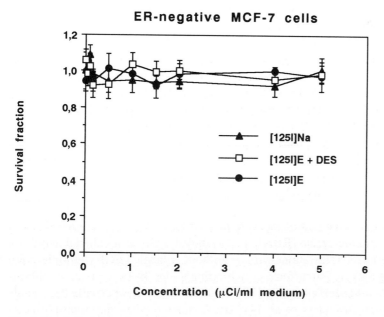

Fig. 2. Graph depicting whether there was cytotoxicity from [125I]E in the presence or absence of the ER competitor DES and cytotoxicity of [125I]Na in ER-rich (Top) and in ER-negative MCF-7 cells (Bottom). Cytotoxicity was evaluated by standard colony-forming assay . Each experiment was performed in triplicate. Values are presented as means ± standard error of the mean (SEM). The formats of the experiments, their protocols and conditions were as described in [30]. Based on the specific activity for [125I]E of ca 2000 Ci/mmol the concentration of 1μCi/ml is equivalent to 0.5 nM iodo-estradiol [33].

with $[^{125}I]E$ could be cytocidal for the ER-rich cancer cells. The second advantage would be that the effect of $[^{125}I]E$ is less cell cycle-dependent and should be effective in ER-rich cells in G_0 because all that should be needed is for the radionuclide-bearing estrogen to bind to the ER-protein and for the (radioligand)-(receptor) complex to form an intimate association with nuclear DNA. The cytotoxicity should not affect cells lacking ER protein or a substantial concentration of ER protein in the nucleus. Disadvantage of this specific ER-selectivity of $[^{125}I]E$ is the selection of ER-negative cells in heterogenic tumors. This may lead to more aggressive, ER-negative tumors resistant to hormone treatment. Up to now there does not exist experimental in vivo data supporting this hypothesis. The site-directed cytotoxicity of an Auger electron-emitting steroid receptor ligand potentially offers a mode for the treatment of steroid receptor-positive cancers. It is not meant to replace the traditional hormonal- or chemotherapies, but should be used as an additive treatment because of the favorable ratio of therapeutic to adverse effects.

Mab 425 (antibody directed against the "epidermal growth factor receptor")

The "epidermal growth factor receptor" (EGF-R) is a 170-kDa glycoprotein which is present in various tissues. It has functional domains, including a cytoplasmic domain of approximately 60 kDa, a short transmembrane sequence and a large 100-kDa external domain [35,36]. The external domain contains the binding site for mitogenic activators, i.e., epidermal growth factor (EGF) and transforming growth factor α (TGFα) [37]. These physiologic ligands are specific for EGF-R. The cytoplasmic domain contains intrinsic EGF-activated tyrosine kinase activity [37].

The clinical importance of the presence of EGF-R in human breast cancer was first reported by Sainsbury et al. [38] and was confirmed by other groups [39–43]. Klijn et al. [7] tried to summarize the results of more than 40 individual groups of investigators. It has been proposed as being an independent prognostic factor determining DFI and OS of individual patients [38–43], an indicator of response to hormonal therapy [7] and, possibly, a target for blocking or cytotoxic agents [44–47]. But according to the review by Klijn et al. [7] there is still controversy about its prognostic value.

The most widely used method of measurement of EGF-R in human breast cancer is the radioligand binding assay (RBA). $[^{125}I]$-labeled EGF is incubated with cell membranes at different concentrations with or without unlabeled EGF (multiple-point radioligand binding assay). The difference in binding under the two sets of conditions described represents the specific binding [38]. Determinations have usually been made by Scatchard analysis. Koenders et al. [41] tried to standardize this multiple-point radioligand binding assay by using a hydroxylapatite (HAP) assay according to recommendations of the EORTC Receptor Study Group. Nicholson et al. [44] simplified the RBA by using a two-point radioligand binding assay. Wyss et al. [48], Bolla et al. [49], and ourselves [50] pointed out that the evaluation of EGF-R content of small tumors with very little tissue available by means of these multiple or two-point radioligand binding assays is sometimes difficult and bears analytical problems. The use of a single-point saturation assay, as described originally by Wyss et al. [48], was therefore recommended and applied by various groups with and without slight modifications [48–50]. The modifications did not alter the results, but facilitated the technical proceeding [50]. Another analytical method, besides these functional RBAs, is the determination of the EGF-R content by means of an immunoenzymetric assay (IEMA) [40] with a specific monoclonal antibody (Mab EGF-R-RI). The immunohistochemical analysis has the advantage of a direct evaluation of the EGF-R distribution and its correlation with morphologic and functional anaplasia in human breast cancer tissue [42,43]. Various antibodies and staining methods have been described [7,42,43,51]. Comparative studies between biochemical and immunohistochemical analyses are very rare [43,51].

Additional, new opportunities for breast cancer therapy are based on the increased biological understanding of tumor growth. As part of the growth factor receptor-mediated cell growth mechanism and the proliferative stimulus of EGF can be inhibited by the use of monoclonal antibodies directed against the external domain of the EGF-R. The monoclonal antibody 425 (Mab 425) binds to a part of the extracellular domain of the EGF-R, which is different from the EGF binding site [52]; Mab 425 does not induce tyrosine kinase activity, but inhibits the binding of EGF and TGFa to the EGF receptor and thus potentially prevents cellular proliferation in vitro [52]. In addition, according to recent evidence, the "antibody dependent cellular cytotoxicity" (ADCC) reaction could mediate the growth inhibitory effect of Mab 425 [45,53].

Table 2. Summary of different dosages applied and different application schedules

Number	Day(s) of application post transplantation	Drug(s) applied	Dosage(s) applied
1	0	Mab 425	2.2 mg i.p.
2	0	Mab 425	1.1 mg i.p.
	0	Mab 425	2.2 mg i.p.
3	0	Mab 425	2.2 mg i.p.
	or 12	Mab 425	2.2 mg i.p.
	or 24	Mab 425	2.2 mg i.p.
4	0	Mab 425	1.1 mg i.p.
	and 12	Mab 425	1.1 mg i.p.
5	1–5	EGF	5 µg/d.
Control	0	PBS	0.45 ml i.p.

Use of Mab 425 in the xenotransplantation model

In the athymic nude mice model with xenotransplanted human carcinomas, the effect of Mab 425 on tumor growth was studied [46,54]. Small pieces (2 × 2 × 2 mm) of five different solid human breast carcinomas and one vulvar epidermoid cancer cell line (A431) were transplanted subcutaneously to the milk line on each side of healthy nude mice (day 0). EGF-R content of these six tumor entities were analyzed biochemically [50]. The five breast cancer tissues were chosen because of their difference in EGF-R content, the results of the analyses range from 0–180 fmol/mg soluble cytosolic protein. The tumor growth in the nude mice was measured by a caliper. A virtual tumor area was calculated by multiplication of the greatest diameter with the perpendicular diameter. Measurements were taken twice a week for the first 3 weeks and then once a week. The mice were treated with different protocols (Table 2).

Tumors with EGF-R concentrations of ≥16 fmol/mg soluble cytosolic protein showed growth inhibition, whereas the growth pattern of EGF-R negative tumors remained unchanged (Fig. 3). Variation of Mab 425 dosage (1.1 vs. 2.2 mg) revealed no difference in the growth inhibitory effect. Different application schedules (application on day 0, 12 or 26) showed different onsets and durations of tumor growth inhibition. Repeated application (1.1 mg, day 0 and 12) was followed by a prolonged inhibitory effect. There was no significant difference in the absolute growth inhibition by Mab 425 after different application dates, suggesting a constant sensitivity to the Mab 425 treatment independent of the tumor size. In all cases the Mab 425 effect was time limited. Tumors regained their proliferative activity after a variable remission interval. Our results suggest that growth inhibition of EGF-R positive tumors by Mab 425 may lead to an additional treatment option for patients with EGF-R positive cancer [46,54].

Clinical phase I/II trail

Due to the promising results from our animal trails and the results from a clinical phase I/II trial with the i.v. application of Mab 425 in patients with EGF-R rich glioblastoma (University Erlangen, Germany), a clinical phase I/II trial for patients with EGF-R rich gynecological or breast cancer was started in the Department of Gynecology and Oncology, University of Frankfurt, Germany. Inclusion criteria were; recurrent or advanced carcinoma of the genital tract or breast, no expected further benefit by established therapy options, Karnofsky index ≥50%, usual laboratory parameters, and EGF-R overexpression in tumor tissue. EGF-R overexpression could be determined by various methods:
1. immunohistochemical detection in frozen tissue [51];
2. single point saturation assay [50];
3. immunoscintigraphy using 1 mg [99]Technecium-labeled Mab 425 [47].
Immunosctigraphy was performed as described previously [47]. [99]Tc is an ideal

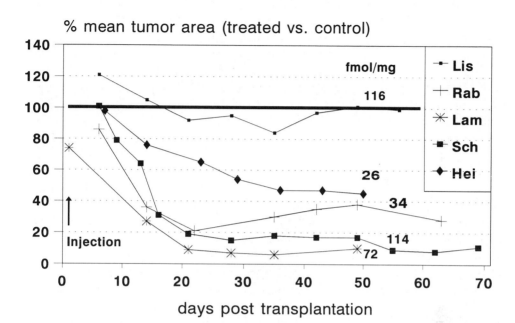

Mab 425 (2.2mg, day 0)
Effects on Solid Tumor Tissue

Fig. 3. Relative tumor growth inhibition of five tumors after Mab 425 application (2.2 mg/ml i.p.). Each curve represents the quotient of the mean tumor area of the treatment vs. the control group. Tumor identification and EGF-R concentrations (fmol/mg soluble protein) are given at the end of each curve [54].

radionuclid for gamma camera imaging (optimum energy peak 140 keV, exclusive gamma radiation emission, half-life of 6 h, cheap). Biodistribution of the ^{99}Tc-labeled Mab 425 was tested in Wistar rats and the "percent injected dose per gram tissue" calculated (Fig. 4). The distribution pattern was typical for a murine antibody. The specific binding of ^{99}Tc-labeled Mab 425 to EGF-R in vivo was tested in the xenotransplantation model. From 3.73 up to 30 μg of ^{99}Tc-Mab 425 was applied i.p. and after 24 h specific binding of EGF-R rich tumors could be detected by scintigraphy. Specific binding could be inhibited by previous application of cold, unlabeled Mab 425 [47].

Up to now, five patients were included: four patients with cervical cancer, of whom two had pelvic relapse including para-aortic positive lymph nodes; two had stage IV disease; one had endometrial cancer. The treatment cycle lasted 4 weeks: daily i.v. infusion of 20 or 100 mg Mab 425 5 days a week, total dose applied 400 or 2,000 mg, respectively. Adverse effects of this treatment possibility were weak, even though the i.v. Mab dose applied was high. Two patients (20 mg per day) showed no adverse effects at all. Fatigue was the main symptom of patients receiving

172

A

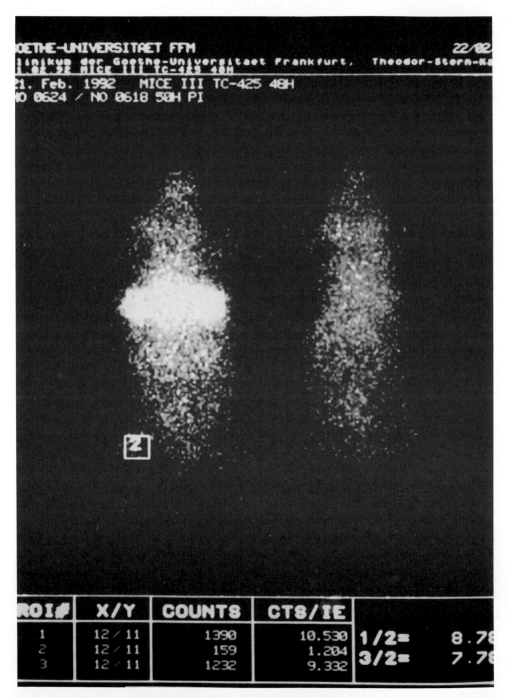

Fig. 4. Scintigraphy of nu/nu mice with ^{99}Tc-labeled Mab 425. A) 30 μg of ^{99}Tc-Mab 425 was applied i.p. and after 24 h specific binding of EGF-R rich tumors could be detected [47].

B

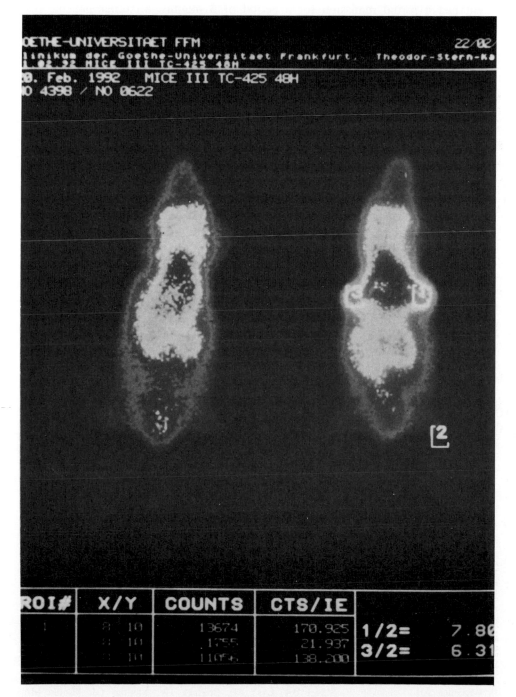

Fig. 4. (continued). B) Specific binding of the ^{99}Tc-labelled Mab 425 could be inhibited by previous application of cold, unlabelled Mab 425 [47].

the higher dose (100 mg per day). One patient developed rosaceiforme acne. One patient had a partial remission for a period of 3 months, all other patients had progression of disease.

Further in vivo investigations are needed to gain insight into the mode of action of Mab 425 in vivo. Tumor characteristics predicting the antibody treatment response are of particular relevance. High histological grade, high proliferative activity (Ki67, S-phase fraction), aneuploidy, negative ER and PR status as well as a high EGF-R content are well known as indicators of poor prognosis [3,7]; further data from ongoing studies will reveal information about the individual tumor sensitivity to Mab 425 treatment. Patients with rapid progression of disease could benefit from this additional treatment option.

Acknowledgements

Studies of [^{123}I]E were supported by DFG-Scha466/1-2 (AS), DFG-Be1215/1-2 (MWB), CA-27476 (JAH), Amersham-Buchler GmbH & CoKG (Dr K.R. Gruner), Braunschweig, Germany, the Deutsche Krebshilfe (KS), and the Mothers' Aid Research Fund of The Chicago Lying-in Hospital. Biomedical Products (Dr R.J. Baranczuk, J.A. Spicer), Overland Park, KS, and The University of Chicago Radiopharmacy (Dr J. Ryan, Erma Peterson) provided [^{125}I]E2 and [^{99m}Tc]rbc gratis, respectively.

Studies of Mab 425 were supported by Grant Be 1068/2-1 from Deutsche Forschungsgemeinschaft, Bonn Germany (HGB), and MERCK Inc., Darmstadt Germany (HGS, HGB) and Paul und Ursula Klein Stiftung, Frankfurt a.M. Germany (MWB). The authors would like to thank the personnel of the Women Clinic's Oncology Laboratory, University of Frankfurt, and especially M. Stegmüller, for their expert help and technical support.

References

1. Scarff RW and Torloni H. Histological typing of breast tumors. International Histological Classification of Tumors No.2, World Health Organization, Geneva, 1968.
2. Early Breast Cancer Trialists Collaborative Group. Lancet 1992;329:1—15.
3. McGuire WL, Clark GM. N Engl J Med 1992;326:1756—1761.
4. Harris JR, Lippman ME, Veronesi U, Willett W. N Engl J Med 1992;327:319—328.
5. Pertschuk LP, Kim DS, Nayer K et al. Cancer 1990;66:1663—1670.
6. Harris JR, Lippman ME, Veronesi U, Willett W. N Engl J Med 1992;327:473—480.
7. Klijn JGM, Berns PMJJ, Schmitz PIM, Foekens JA. Endocrine Rev 1992;13:3—17.
8. Klijn JGM, Berns PMJJ, van Putten WLJ, Bontenbal M, Alexieva-Figusch J, Foekens JA. In: Senn HJ, Gelber RD, Goldhirsch, Thürlimann B (eds) Adjuvant Therapy of Breast Cancer IV. Recent Results in Cancer Research 127. Berlin, Germany: Springer Verlag, 1993;77—88.
9. Ozzello L, DeRosa C, Habif DV, Greene GL. Cancer 1990;67:455—462.
10. McGuire AH, Dehdashti F, Siegel BA, Lyss AP, Brodack JW, Mathias CJ, Mintun MA, Katzenellenbogen JA, Welch MJ. J Nucl Med 1991;32:1526—1532.
11. Scheidhauer K, Müller S, Smolarz K, Bräutigam P, Briele B. Nucl Med 1991;30:84—99.

12. Schober O, Scheidhauer K, Jackisch C, Schicha H, Smolarz K, Bolte A, Reiners C, Höffken K, Biersack HJ, Briele B, Bräutigam P. Lancet 1990;335:1522–1523.
13. Anderson RE, Holt JA. Gynecol Oncol 1989;34:80–83.
14. Seevers RH Jr In: Spencer RP, Seevers RH Jr, Friedman AM (eds) Radionuclides in Therapy. Boca Raton: FL:CRC Press, 1987;145–166.
15. Holt JA, Artwohl JA, Mercer LJ, Pryde PG. Int J Radiat Oncol Biol Phys 1991;20:567–573
16. Scharl A, Beckmann MW, Artwohl JA, Kullander S, Holt JA. Int J Radiat Oncol Biol Phys 1991; 21:1235–1240.
17. Scharl A, Kullander S, Beckmann MW, Spicer JA, Baranczuk RJ, Holt JA. Am J Obstet Gynecol 1991;165:1847–1853.
18. Scharl A, Holt JA. Int J Radiat Oncol Biol Phys 1993;26:285–290.
19. Artwohl JA, Scharl A, Beckmann MW, Holt JA. Contemp Top Lab Animal Sci 1993;32:13–16.
20. Ritschel WA, Banerjee PS. Meth Exp Clin Pharmacol 1986;8:603–610.
21. Lobo RA. Obstet Gynecol Clin North Am 1987;14:143–150.
22. Lobo RA, Cassidenti DL. J Reprod Med 1992;37:77–82.
23. Holt JA, Scharl A, Kullander S, Beckmann MW. Acta Obstet Gynecol Scand 1992;71:39–54.
24. McLaughlin WH, Milius RA, Pillai KMR, Edasery JP, Blumenthal RD, Bloomer WD. J Natl Cancer Inst 1991;81:437–440.
25. Bloomer WD, McLaughlin WH, Milius RA, Weichselbaum RR, Adelstein SJ. J Cell Biochem 1983; 21:39–45.
26. McManaway ME, Jagoda EM, Kasid A, Eckelman WC, Francis BE, Larson SM, Gibson RE, Reba RC, Lippman ME. Cancer Research 1987;47:2945–2949.
27. Woo DV, Li D, Mattis JA, Steplewski Z. Cancer Res 1989;49:2952–2958.
28. Baranowska-Kortylewicz J, Makrigiorgos GM, Van den Abbeele AD, Berman RM, Adelstein SJ. Int J Radiat Oncol Biol Phys 1991;21:1541–1554.
29. Adelstein SJ, Kassis AI. Nucl Med Biol 1987;3:165–169.
30. Rao DV, Narra VR, Howell RW, Govelitz GF, Sastry KSR. Lancet 1989;2:650–653.
31. Kassis AI, Sastry KSR, Adelstein SJ. Radiat Res 1987;109:78–89.
32. DeSombre ER, Harper PV, Hughes A, Mease RC, Gatley SJ, DeJesus OT, Schwartz JL. Cancer Res 1989;48:5805–5809.
33. Beckmann MW, Scharl A, Rosinsky BJ, Holt JA. J Cancer Res Clin Oncol 1993;119:207–214.
34. Scharl A, Beckmann MW, Schreiber JR, Hol JA. Onkologie 1991;14:433–436.
35. Downward J, Yarden Y, Mayes E et al. Nature 1984;307:521–527.
36. Niederacher D, Beckmann MW, Scharl A, Picard F, Schnürch HG, Bender HG. In: Koldovsky U, Kreienberg R (eds) Tumorimmunologie in der Gynäkologie. Aktuelle Onkologie 1994;79:73–79.
37. Carpenter G. Ann Rev Biochem 1987;56:881–914.
38. Sainsbury JRC, Malcolm AJ, Appleton DR, Farndon JR, Harris AL. J Clin Pathol 1985;38: 1225–1231.
39. Sainsbury JRC, Farndon JR, Needham GK, Malcolm AJ, Harris AL. Lancet 1987;1:1398–1402.
40. Grimaux M, Mady E, Remvikos Y, Laine-Bidron C, Magdelenat H. Int J Cancer 1990;45:255–262.
41. Koenders PG, Farverly D, Beex LVAM, Bruggink EDM, Kienhuis CBM, Benraad TJ. Eur J Cancer 1992;28:693–697.
42. Scharl A, Göhring UJ, Vierbuchen M, Küsters B, Würz H. Geburtsh u Frauenheilk 1990;50:877–882.
43. Toi M, Hamada Y, Nakamura T et al. Int J Cancer 1989;43:220–225.
44. Nicholson S, Sainsbury JRC, Needham GK, Chambers P, Farndon JR, Harris AL. Int J Cancer 1988;42:36–41.
45. Rodeck U, Herlyn M, Herlyn D et al. Cancer Res 1987;47:3692–3696.
46. Schnürch HG, Beckmann MW, Stegmüller M, Bender HG. In: Klapdor R (ed). Tumor Associated Antigens, Oncogenes, Receptors, Cytokines in Tumor Diagnosis and Therapy at the Beginning of the 90th. München, Germany: Zuckschwerdt Verlag, 1992;621–624.
47. Beckmann MW, Niederacher D, Schnürch HG, Baum RP, Adam S and Bender HG. In: Koldovsky U, Kreienberg R (eds) Tumorimmunologie in der Gynäkologie. Aktuelle Onkologie 1994;79:58–72.

48. Wyss R, Fabbro D, Regazzi R, Borner C, Takahashi A, Eppenberger U. Anticancer Res 1987;7: 721—728.
49. Bolla M, Chedin M, Souvignet C, Marron J, Arnould C, Chambaz E. Breast Cancer Res Treat 1990; 16:97—102.
50. Beckmann MW, Stegmüller M, Niederacher D, Tutschek B, Schnürch HG,Bender HG. Tumordiagn Ther 1993;14:41—48.
51. Beckmann MW, Tutschek B, Krüger KH, Niederacher D, Risse BC, Ruppert C, Schnürch HG, Bender HG. Int J Oncol 1993;3:389—397.
52. Rodeck U, Williams N, Murthy U, Herlyn M. J Cell Biochem 1990;44:69—79.
53. Sutter A, Hekmat A, Luckenbach GA. Pathobiology 1991;59:254—258.
54. Schnürch HG, Stegmüller M, Vering A, Beckmann MW, Bender HG. Eur J Cancer 1994.

Integration of prognostic factors: can we predict breast cancer recurrences?

Gary M. Clark and Susan G. Hilsenbeck

University of Texas Health Science Center at San Antonio, Division of Medical Oncology, 7703 Floyd Curl Drive, San Antonio, TX 78284-7884, USA

Abstract. New potential prognostic factors for patients with breast cancer are being discovered everyday, and it is becoming increasingly more difficult for physicians and patients to integrate the information provided by these factors. Several multivariate techniques that can and have been applied to this problem will be reviewed, including Cox's proportional hazards model, recursive partitioning, correspondence analysis, and neural networks. Examples from the literature are given to demonstrate how these techniques have been applied to predict breast cancer recurrences.

Introduction

With the proliferation of potential prognostic factors for breast cancer, it is becoming increasingly more difficult for physicians and patients to integrate the information provided by these factors into an accurate prediction of clinical outcome. As we measure more and more factors, the chances increase that the factors will provide conflicting information. Some will be associated with a good prognosis, but others will be indicative of a poor prognosis. How should we weight these factors and combine their values to produce a global assessment of likely clinical outcome for an individual patient?

As we consider this problem it is necessary to keep in perspective the reasons for measuring prognostic factors in the first place. We have previously described three different situations where prognostic markers might be useful for primary breast cancer [1]:

— To identify patients whose prognosis is so good following local surgery that the addition of systemic adjuvant therapy would not be cost beneficial.
— To identify patients whose prognosis is so poor with conventional treatment that other forms of more aggressive therapy might be warranted.
— To identify patients who are or are not likely to benefit from specific therapies.

Although the objectives are quite different, the statistical and mathematical strategies that can be used to address these situations may be quite similar.

Modeling approaches

There are a variety of methods for combining factors into a composite index. The TNM and UICC classifications are examples that combine tumor size, axillary lymph node status and presence of metastatic disease into simple staging systems. However, as more factors become available we should consider more sophisticated methods for combining information from multiple factors.

The traditional method for performing multivariate statistical analyses is to create Cox nonproportional hazards models. Recently, other multivariate strategies have been applied to prognostic factor problems, including recursive partitioning, correspondence analysis, and neural networks. We will briefly review each of these techniques and describe some clinical applications where they have been used.

Cox proportional hazards model

The Cox proportional hazards model has become a standard statistical technique for performing multivariate survival analyses [2]. It is a very general model that describes the hazard function (instantaneous force of mortality) at time t for an individual with covariates $\underline{z} = (z_1,...,z_p)$ as:

$$h(t,\underline{z}) = H(t) \exp[\underline{B}'\underline{z}]$$

where H(t) is an arbitrary baseline hazard function for an individual with covariate vector $\underline{z} = (0,...,0)$ and $\underline{B} = (B_1,...,B_p)$ is the vector of regression coefficients. Since H(t) is completely unspecified, this model has been described as "partially nonparametric". Estimation of the regression coefficients is usually accomplished by an iterative procedure based on partial likelihood functions and hypothesis testing relies on assumptions of asymptotic normality.

A simple form of a Cox model is the Nottingham Prognostic Index [3] for patients with primary breast cancer:

Index = 0.2 * Tumor Size + Stage + Grade

This index is essentially the exponential component, $\exp[\underline{B}'\underline{z}]$, of the hazard function from a Cox model. This index has been used to divide patients into one of three prognostic groups: good, moderate and poor as shown in Table 1.

A recent update [4] based on 1629 cases of operable breast cancer, demonstrated

Table 1. Prognostic groups defined by the Nottingham Prognostic Index

Group	Prognostic index	% of patients	15-year survival
Good	<3.4	29%	80%
Moderate	3.4–5.4	54%	42%
Poor	>5.4	17%	13%

that these three groups have quite different 15-year survival rates: 80% for the good group, 42% for the moderate group and 13% for the poor group. An age-matched female control population had an 83% 15-year survival, suggesting that the good group of patients would probably have gained very little benefit from the addition of adjuvant therapy. One criticism of the Nottingham Prognostic Index is that it includes the somewhat subjective evaluation of histologic grade. Although reports of poor reproducibility of nuclear and histological grade have dampened the enthusiasm of some clinicians, with more standardized grading systems and continued education of pathologists trained in breast cancer histology, these factors could be among the most useful predictors of clinical outcome. The success of this index suggests that new markers, in combination with standard factors, could provide an even more accurate and more powerful mechanism for predicting the likely clinical outcome of breast cancer patients.

However, as the pendulum swings towards administration of systemic adjuvant therapy for all patients with primary breast cancer, the window of opportunity for identifying patients who do not require systemic adjuvant therapy becomes smaller and smaller. Even though the results of the recent overview analysis of breast cancer clinical trials [5] demonstrated that the actuarial 10-year survival of patients in the control arms of studies evaluating endocrine therapy was 71%, it is now almost impossible to initiate a randomized clinical trial with an untreated control arm.

Since potential prognostic factors are often dichotomized before constructing Cox models (e.g., high vs. low, present vs. absent, treatment vs. no treatment), there is the perception among some people that this is the preferred method to represent factors. However, Kalbfleisch and Prentice [6] note that the survival curve for one level of a binary factor will necessarily dominate the other and they caution about over-interpretation of the results from such models. The Cox model is designed to include variables measured on a continuum and can easily incorporate interactions between factors.

An example of the use of a Cox model that demonstrated the power that can be attained by more optimal representation of the prognostic factors was given by Clark et al. [7]. The objective of that study was to identify patients with primary breast cancer whose prognosis was so poor with conventional adjuvant therapy that other forms of more aggressive therapy might be warranted. The data set consisted of patients with node-positive breast cancer. Two different Cox models were constructed: one using only dichotomous factors and another using a combination of dichotomous and continuous factors. A total of 21% of the patients had 10 or more positive nodes, a standard eligibility criterion for ongoing trials of high dose chemotherapy followed by bone marrow transplantation. The dichotomous Cox model identified a subset of 27% of the patients whose 4-year survival was equal to or worse than that of the patients with 10 or more positive nodes (Table 2). By using appropriate choices of data transformations for inherently continuous factors, a subset comprised of 49% of the patients was identified with 4-year survival equal to that of the previous two subsets.

Another misconception about the Cox model is that it cannot provide probability

180

Table 2. Percentage of patients with 4-year survival worse than 40%[a]

Model	%
Only dichotomous factors	27%
Dichotomous and continuous factors	49%

[a]4-year survival of patients with 10 or more positive nodes was 40%.

estimates for individual patients. The difficulty is that the arbitrary baseline hazard function must first be estimated. However, standard statistical packages such as SAS and BMDP do include options for estimating this hazard function and therefore for obtaining survival probability estimates for patients with a given set of covariates.

There are two basic assumptions for the Cox model:
— the ratio of the hazard functions for two individuals with different sets of covariates does not depend on time (proportional hazards); and
— the effect of covariates on the hazard function is log-linear.

Several authors have proposed formal tests of the proportional hazards assumption [6,8—10] and procedures now exist for incorporating time-dependent covariates. Hastie and Tibshirani [11] have proposed a method based on additive nonparametrically smoothed functions of the covariates within the proportional hazards framework to overcome violations of the log-linear assumption, and Hastie et al. [12] demonstrated the technique using data from a breast cancer clinical trial. This approach is very appealing for data sets with nonlinear interactive effects, but has not yet been widely applied to clinical data. Frequently, definitive Cox modeling is preceded by an empirical search for appropriate transformations of the covariates (e.g., logarithms, square roots, coding systems, etc.). With Hastie and Tibshirani's approach it may be possible to merge these steps. Inclusion of stratification factors, such that the cases within each stratum conform to the proportional hazards assumption, can also address departures from proportionality.

When evaluating potential prognostic factors, one is often confronted with a large panel of candidate factors and asked to determine which combination of factors is "best" for predicting the clinical outcome of patients. The immediate questions that must be addressed are:
— how to represent the factors, (e.g., raw or transformed data, continuous or categorical representations, fixed or time-dependent covariates);
— which (if any) interactions should be included; and
— how to select factors for the final model.

The most commonly used variable selection technique is a stepwise procedure. The forward selection method builds a model by adding prognostic variables one-by-one using a stepwise approach. Starting with a model that contains no covariates, a factor is added to the model if it significantly improves the ability of the model to predict clinical outcome. Significance is defined by the large-sample partial likelihood ratio test. The backward elimination method begins with all variables in the model and then eliminates nonsignificant variables one-by-one. Because of the correlations that exist among potential prognostic factors, neither stepwise procedure is guaranteed to result in an optimal model. More recently, algorithms have been developed to examine all possible subsets of prognostic factors in a computationally efficient manner. Chapman et al. [13] have compared all-subsets and stepwise models using a node-positive breast cancer example, and offer advice about data representation and model construction.

Due to limited sample sizes, a Cox model is usually constructed using data from all available patients. As in all modeling problems there is a tendency to over-fit the model. It performs very well on the patients used to build the model, but unless the model is validated on an independent set of patients its ability to generalize to other patient populations is questionable.

A limitation of Cox's proportional hazards model is the need to have each covariate measured for each patient. Although techniques exist to estimate missing data values, these approaches involve additional assumptions that often cannot be substantiated. Alternative approaches, such as recursive partitioning, provide capabilities for analyzing data sets with missing values without the need for large scale data estimation.

Recursive partitioning

During the last decade, tree-structured regression methods, known generically as either classification and regression trees (CART) or recursive partitioning, have been developed as alternatives to classical regression and discriminant analysis [14]. Tree-structured regression procedures typically result in the construction of a binary classification tree, in which cases are separated (split) based on values of covariates (or combinations of covariates). The analytic algorithm involves growing trees by splitting, followed by tree pruning and terminal node amalgamation. The ability of the tree to correctly classify cases is evaluated based on measures of cost-complexity or misclassification rates. Evaluation of tree performance often involves cross-validation or other resampling/bootstrapping procedures.

Recently, these methods have been extended to the analysis of survival data with censored observations [15–20]. As before, regression trees are grown as a series of nodes. The initial or root node consists of the entire sample. For each node, every possible split (i.e., dichotomization into classes of low and high values for ordinal factors, or any disjoint subsets for nominal factors) for each predictor variable is examined, and the best split is selected. "Best" is defined in terms of a goodness-of-split criterion. Segal [16] has proposed that this criterion be any member of the Tarone-Ware or Harrington-Fleming classes of test statistics. These classes include the logrank statistic and the generalized Wilcoxon statistic that are commonly used in survival analyses. Split criteria based on Martingale residuals (a measure of minimization of excess expected deaths) and the Wasserstein L_p metric (a measure of distance between distributions) have also been used. The splitting process is continued until no further splits are possible. The tree is then pruned by eliminating noninformative nodes, and nodes with similar clinical outcomes are amalgamated to

Table 3. Subgroups formed by recursive partitioning

Subgroup	N	6-year DFS
N_0, ER+	56	95%
N_{13}, srp27+	31	55%
N_{13}, srp27– or N_0, ER–	81	70%
N_{4+}	45	30%

produce the final regression tree.

An example of a recursive partitioning analyses was published by Thor et al. [21] who evaluated the prognostic significance of stress response shock protein srp-27 in a series of 213 breast cancer patients. Also included in the analysis were the number of positive lymph nodes (N), estrogen receptor status (ER) and tumor size (T stage). This analysis produced 4 subgroups of patients with different 6-year disease-free survival outcomes as shown in Table 3.

A Cox model was also created using the same patients and the same factors. As displayed in Table 4, this model yielded somewhat different results.

The authors concluded that srp27 was a significant predictor of clinical outcome, but only for patients with 1 to 3 positive lymph nodes. This finding would have been missed if only the Cox analysis had been performed.

Regression trees provide an intuitively attractive alternative to Cox proportional hazards modeling for the identification of prognostically important subgroups. Trees are easy to use in clinical practice and could help identify small groups of patients with high or low risks of recurrence, or complex covariate interactions that are only exhibited in certain subgroups.

Preliminary work with several real datasets emphasizes three principles:
— the method is potentially useful in examining previously unknown interactions in subgroups;
— the method requires careful application on mature, complete datasets; and
— validation of the final tree is required before results can be generalized.

Performance may degrade in the presence of high rates of censoring, significant departures from hazard proportionality among adjacent nodes, and prognostically unimportant covariates. In particular, simulation studies by LeBlanc and Crowley [19] suggest that spurious end-cut splitting is a problem in datasets lacking true factor effects and that the performance of different split selection algorithms depends on data structure and censoring patterns. This emphasizes the need for screening and

Table 4. Cox model

Factor	p
Node positivity	<0.001
T stage	0.003
ER positivity	0.074
srp27 positivity	0.242

exploratory analyses prior to the construction of prognostic models. Based on our pilot studies and discussions with other investigators, classification trees can be unstable in topology and may be undesirably sensitive to other cases. Therefore we view recursive partitioning as an exploratory adjunct to other methods and a useful form for data summarization, rather than as a definitive analytic method. Confirmatory procedures, such as cross-validation resampling or bootstrapping, must be key features of any analysis. Comparative results for several splitting strategies could provide a means to assess topological stability. The optimal selection of splitting, pruning and amalgamation strategies and the assessment of tree performance remain issues of theoretical and practical investigation.

Correspondence analysis

Correspondence analysis is a form of factor analysis for categorical variables that can be used to extract new underlying dimensions that combine the information contained in the individual variables into a more compact, nonredundant form. It was first described in 1933 [22], but seems to have been invented independently by several different investigators. As described in the SAS manual [23], it is known by different names throughout the world: optimal scaling, reciprocal averaging, optimal scoring and appropriate scoring in the United States, quantification method in Japan, homogeneity analysis in the Netherlands, dual scaling in Canada and scalogram analysis in Israel. The results of correspondence analyses can be used to graphically examine associations between categories, and can also be used to graphically represent individual patients.

Hilsenbeck et al. [24] analyzed 354 breast cancer patients with data on standard clinical variables and some new investigational factors. Correspondence analysis was used to extract new dimensions that contained integrated information from these multiple factors. After plotting these dimensions together with recurrence status and examining relationships between categories, the dimensions were then entered into traditional Cox model analyses to produce predictive models. This preliminary study suggested that correspondence analysis can be applied to the evaluation of breast cancer prognostic factors, providing insight into the complex interrelationships of prognostic factors and simplifying their interpretation.

Neural networks

Neural networks were originally developed as pattern recognition systems in the 1960s [25,26]. However, they were not widely used until the 1980s when an improved learning technique, the back-propagation of errors algorithm, led to improved performance [27]. Neural networks are self-organizing systems that are capable of learning the significant features of data sets. They can be "trained" using a training data set to predict the outcome or class of an event after being presented with a set of input variables. During training, their internal structures change in such a way as to reinforce correct decisions and make incorrect responses less likely.

Table 5. Steps for performing neural network analysis on censored survival data

1. Data examination, transformation, and normalization.
2. Preparation of training and validation sets.
3. Coding of time as a prognostic variable.
4. Training of the neural network.
5. Examination of neural network function.
6. Use of neural network as a predictive model.

Neural networks have been used in such diverse problems as predicting protein secondary structure, detecting promoter regions in DNA sequences, synthesizing spoken language, diagnosing and treating hypertension and identifying military targets based on Sonar signatures. Ravdin et al. [28,29] have recently shown that they can also be used to predict the probable clinical course of breast cancer patients and listed 6 steps for performing neural network analysis on censored survival data (Table 5).

The most commonly used networks consist of three layers of neurons or units: an input layer that accepts external information; a hidden layer that is crucial to the information processing; and an output layer where further information processing takes place and the system response is calculated. During training the connections between units are gradually modified in order to maximize correct responses and minimize incorrect responses. The system responds to an input with a numerical output which is then compared with the known or target output. The mathematical technique that is used to change the connection weights is known as the "back-propagation of errors algorithm." This algorithm allows neural networks to efficiently search multidimensional variable spaces for optimal solutions to very complex problems.

Neural networks have the potential to over-train (i.e., memorize the input variables for each patient) on training data sets. Therefore it is imperative that independent training and test sets of patients be available for the construction of a neural network. During training there are two competing processes:
— learning of general features in the data set that are predictive of outcome, and
— gradual learning of specific features of individual data points.
Both of these processes occur simultaneously with a gradual improvement in the learning of general features, but reaching an optimal level and then usually degrading as the second over-training process becomes stronger. Network evolution is monitored by periodically interrupting training to test the performance of the network on an independent set of patients. A plot of accuracy against a number of training iterations usually results in a smooth increase to a maximum followed by a stabilization or decline. By saving the intermediate network structures during training and testing, one can obtain the "optimal" network for further evaluation.

One of the criticisms of neural networks is that they are like a "black box", where data enter and predictions emerge, but the process inside is unknown and unknownable. However, Ravdin and Clark [29] describe techniques to examine the structure of a neural network and determine how it uses the prognostic data. They also present a method for obtaining estimates of survival probabilities.

Summary

Several techniques are available for integrating information provided by multiple prognostic factors. However, very few comparative studies have been performed to determine which methods might be optimal under different conditions. Is it possible to predict breast cancer recurrences with currently available information? The Nottingham Prognostic Index [3,4] appears to provide excellent separation of patients into subsets with different prognoses. The prognosis of the Good group appears to be as good as that of an age-matched cohort of women without breast cancer. If these results could be confirmed by an independent group of investigators, this index, or a refinement of this index, could be used to identify a subset of women who would have little chance of benefiting from systemic adjuvant therapy. Cox models that identify twice as many poor risk patients with the same clinical outcome as patients with 10 or more positive nodes [7] could be used to increase enrollment on clinical trials to evaluate the efficacy of high dose chemotherapy.

However, future progress in the integration of prognostic factors will require large data sets with long-term follow-up of patients in which the majority of the standard and new prognostic factors are assessed together. To address the third area described in the Introduction, the identification of patients who are or are not likely to benefit from specific therapies, coordination with large clinical trials with standardized therapies will be required. And finally, new procedures must be developed to compare the performance of the various statistical techniques when they are applied to the same data sets.

Acknowledgements

Supported in part by grants CA30195 and CA54174 from the National Cancer Institute.

References

1. Clark GM. Do we really need prognostic factors for breast cancer. Breast Cancer Res Treat (in press).
2. Cox DR. J R Stat Soc (B) 1972;34:187–202.
3. Haybittle JL, Blamey RW, Elston CW, Johnson J, Doyle PJ, Campbell FC, Nicholson RI, Griffiths K. Br J Cancer 1982;45:361–366.
4. Galea MH, Blamey RW, Elston CE, Ellis IO. Breast Cancer Res Treat 1992;22:207–219.
5. Early Breast Cancer Trialists' Collaborative Group. Lancet 1992;339:1–15,71–85.
6. Kalbfleisch JD, Prentice RL. In: The Statistical Analysis of Failure Time Data. New York: John Wiley & Sons, 1980.

7. Clark GM, Wenger CR, Beardslee S, Owens MA, Pounds G, Oldaker T, Vendely P, Pandian MR, Harrington D, McGuire WL. Cancer 1993;71:2157–2162.
8. Mitchell DM, Spitz PW, Young DY, Bloch DA, McShane DJ, Fries JF. Arthritis Rheum 1986;29: 706–714.
9. Crowley J, Hu M. J Am Stat Assoc 1977;72:27–36.
10. Arjas E. J Am Stat Assoc 1988;83:204–212.
11. Hastie T, Tibshirani R. Biometrics 1990;46:1005–1016.
12. Hastie T, Sleeper L, Tishirani R. Breast Cancer Res Treat 1992;22:241–250.
13. Chapman JAW, Trudeau ME, Pritchard KI, Sawka CA, Mobbs BG, Hanna WM, Kahn H, McCready DR, Lickley LA. Breast Cancer Res Treat 1992;22:263–272.
14. Brieman L, Friedman JH, Olshen RA, Stone C. In: Classification and Regression Trees. Belmont, CA: Wadsworth International Group, 1984.
15. Gordon L, Olshen RA. Cancer Treat Rep 1985;69:1065–1068.
16. Segal MR. Biometrics 1988;44:35–47.
17. Ciampi A, Chang CH, Hoss S, McKinney S. In: Umphrey G (ed) Proceedings from Joshi Festschrift. Amsterdam: North-Holland, 1987;23–50.
18. Ciampi A, Lawless JF, McKinney SM, Singhal K. J Clin Epidemiol 1988;8:737–748.
19. LeBlanc M, Crowley J. Biometrics 1992;48:411–425.
20. Segal MR, Bloch DA. Stat Med 1990;8:539–550.
21. Thor A, Benz C, Moore D II, Goldman E, Edgerton S, Landry J, Schwartz L, Mayall B, Hickey E, Weber LA. J Natl Cancer Inst 1991;83:170–178.
22. Richardson M, Kuder GF. Personnel J 1933;12:36–40.
23. SAS/STAT User's Guide, Version 6, 4th edn. Cary, NC: SAS Institute, Inc., 1990;615–675.
24. Hilsenbeck SG, Clark GM, Chamness G, Osborne CK. Proc Am Assoc Cancer Res 1993;34:193.
25. Rosenblatt T. In: Principles of Neurodynamics. New York: Spartan Books, 1969.
26. Minsky ML, Papert S. In: Perceptrons. Cambridge, MA: MIT Press, 1969.
27. Rumelhart DE, Hinton GE, Williams RJ. Nature 1986;323:533–536.
28. Ravdin PM, Clark GM, Hilsenbeck SG, Owens MA, Vendely P, Pandian MR, McGuire WL. Breast Cancer Res Treat 1992;21:47–53.
29. Ravdin PM, Clark GM. Breast Cancer Res Treat 1992;22:285–93.

Enzyme-linked immunosorbent assay for plasminogen activator inhibitor-1 (PAI-1)

Michael D. Kramer*, Jutta Link, Jeannette Reinartz and Frank Buessecker

Institut für Immunologie und Serologie der Universität, Laboratorium für Immunpathologie, Im Neuenheimer Feld 305, D-69120 Heidelberg, Germany

Abstract. Reported here is an enzyme-linked immunosorbent assay (ELISA) for the quantification of the plasminogen activator inhibitor-1 (PAI-1). The assay was developed to detect the total amount of PAI-1 in cell culture supernatants. Besides PAI-1, cultured cells can also produce two types of plasminogen activators (PA) (i.e., urokinase-type PA (uPA) and tissue-type PA (tPA)). The PAs can be inhibited by the PAI-1 via formation of covalent enzyme/inhibitor complexes. Thus the PAI-1 may be present in cell culture supernatants as free inhibitor or as complex with either type of PA. Monoclonal antibodies that recognized the free as well as the complexed PAI-1 were selected for construction of the ELISA. The ELISA test was found to be reliable, easy to perform and to permit the detection of PAI-1 in serum-free and serum-containing cell culture supernatants. The test was used to analyze PAI-1 in culture supernatants of a PAI-1-producing cell line.

Introduction

Plasmin obtained via activation of the ubiquitous proenzyme plasminogen is thought to play an important role in the pericellular proteolysis [1]. Plasminogen activation is performed by specialized serine proteinases, the so-called plasminogen activators (PA) [2–4]. Two types of PA that can be produced by a variety of cell types [2] are known: urokinase-type PA (uPA), and tissue-type PA (tPA). PAs are regulated by so-called plasminogen activator inhibitors (PAI). In humans there are two different types of PAIs, PAI-1 and PAI-2. They are different gene products and have distinct immunological and biochemical properties [5]. Both inhibitors inactivate uPA and tPA, their function under physiological conditions seems to be distinct: PAI-1 is the principal PAI in normal human plasma, whereas PAI-2 is present in plasma during late pregnancy only. In vitro both types of PAI can be produced by a variety of cell types [5]. Since both types of PAI inhibit uPA and tPA, functional determination of the inhibitors cannot distinguish between the presence of PAI-1 or PAI-2. Thus, the selective detection and quantification of PAI-1 or PAI-2 requires immunological methods.

Address for correspondence: Michael D. Kramer, Institut für Immunologie und Serologie der Universität, Laboratorium für Immunpathologie, Im Neuenheimer Feld 305, D-69120 Heidelberg, Germany. Tel.: +49-6221-564008. Fax: +49-6221-564030.

Here we describe an enzyme-linked immunosorbent assay (ELISA) test for the quantification of PAI-1. The influence on the assay performance of fetal calf serum (FCS) — often used as addition to cell culture media — was studied. Furthermore, the assay system was used to analyze the conditioned media of the PAI-1-producing cell line HT-1080.

Materials and Methods

Materials

High molecular weight uPA (HMW-uPA) was obtained from Serono (Freiburg). Single chain tPA was from Kabi Vitrum, purchased via Pharmacia (Freiburg); two-chain tPA was obtained from Paesel und Lorei (Frankfurt). Human PAI-1 was from American Diagnostica, purchased via Ortho (Neckargemünd). Human vitronectin was purchased from Telios (USA).

Poly- and monoclonal antibodies

Polyclonal goat anti-PAI-1 IgG (no. 395 G) was obtained from American Diagnostica, peroxidase-labeled goat-anti-mouse IgG Fc antibodies (no. 115-035-071) were from Dianova (Hamburg). Monoclonal anti-PAI-1 antibodies were raised according to established methods [6,7] after immunization with human natural PAI-1.

Cell culture and preparation of serum-free culture supernatants

The human fibrosarcoma cell line HT-1080 [8] was cultivated in RPMI 1640 medium supplemented with 5% FCS and 2 mM L-glutamine. Conditioned media were harvested after different intervals of time by removing the cells by centrifugation (1,000 × g, 10 min). Adherent HT-1080 cells were detached by trypsinization (0.25%), washed twice and seeded at a density of 1×10^5 cells/ml into 6-well plates in the serum-free medium "HL-1". Supernatants collected after 24, 48, 72 and 96 h of incubation were centrifuged for 10 min at 10,000 rpm and stored at −80°C until analysis. The HL-1 medium interferes with neither the immunological nor the functional detection of PAs or PAIs.

Preparation of PAI/PA and PAI/vitronectin complexes

PAI-1/PA or PAI-1/vitronectin complexes were generated by incubating PAI-1 (10 ng/ml) with 50, 100, 200 ng/ml two chain tPA (tc-tPA), with 25, 50, 200 ng/ml HMW-uPA or with 80, 320, 640 ng/ml vitronectin for 45 min at room temperature in phosphate-buffered saline (PBS) [9]. Formation of complexes was tested by specific "complex-ELISAs". In brief: microtiter plates were coated with a polyclonal goat-anti-PAI-1 IgG. After blocking of nonspecific binding sites and subsequent

Table 1. Verification of the formation of PAI-1/PA or PAI-1/vitronectin complexes by an ELISA that specifically detects such complexes

Preparations tested in the "complex-ELISA"[a]	Signal obtained (^{A}490-450 nm)
PAI-1/uPA	0.42
PAI-1/tPA	0.45
PAI-1/vitronectin	0.762

[a]"complex-ELISA" (refer to Materials and Methods for details of the method). The formation of complexes is indicated by the positive signals, when the respective PAI was pre-incubated with either uPA, tPA or vitronectin. The non-complexed enzymes (uPA, tPA), or the free PAI-1 and free vitronectin did not yield positive signals (data not shown).

incubation with complex preparations, monoclonal antibodies with specificity for uPA, tPA or vitronectin were added. The further performance was similar to the protocol of the PAI-1-specific ELISA described below. The results obtained with the "complex-ELISAs" are given in Table 1.

Enzyme-linked immunosorbent assay for quantification of PAI-1

Flat bottom microtiter plates (no. 469914; Nunc, Wiesbaden) were coated with 100 µl/well of polyclonal goat-anti-PAI-1 IgG (1 µg/ml) in 50 mM Na_2HCO_3, 3 mM NaN_3, pH 9.6 for 18 h at 4°C. Nonspecific binding sites were blocked with 200 µl/well of PBS, 0.2% (w/v) gelatine (Merck, Darmstadt) for 1 h at room temperature with continuous shaking. The precoated plates were then washed 2 × with 200 µl/well

Table 2. Characteristics of the PAI-1-specific ELISA

Parameter	PAI-1 ELISA
monoclonal antibody	HD-PAI-1 14.1
reactivity with	PAI-1 PAI-1/uPA-complex PAI-1/tPA-complex PAI-1/vitronectin-complex
sensitivity	0.15 ng/ml
precision (C.V.[b] < 10%)	>1 ng/ml
working range	0.2—10 ng/ml
linearity (r^a)[c]	0.988—0.999
interassay variability	11.8—13.5 %

[a]coefficient of correlation; [c]C.V. = coefficient of variation; [c]linearity was determined within the given working range.

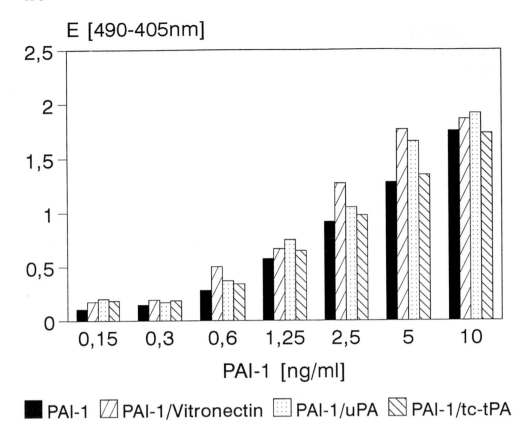

Fig. 1. Detection of free PAI-1 and PAI-1/vitronectin-, PAI-1/uPA- and PAI-1/tPA-complexes in the PAI-1-specific ELISA. The mean extinction value of each bar was calculated from the determination of four distinct complex-preparations.

of PBS containing 0.05% Tween 20 (Merck, Darmstadt) (PBS/Tw). The test samples, diluted in PBS/Tw, were added (100 µl/well) and incubated for 1 h at room temperature with continuous shaking. After the plates had been washed 4 × using PBS/Tw, the mAb HD-PAI-1 14.1 diluted to 1 µg/ml was added (100 µl/well) and incubated for 1 h at room temperature with permanent shaking. The plates were washed 4 × using PBS/Tw, peroxidase-labeled goat-antimouse IgG Fc (Dianova GmbH; Hamburg) diluted 1/5000 in PBS/Tw (100 µl/well) was added and the plates were incubated for 1 h at room temperature. After washing the plates 4 × using PBS/Tw, bound peroxidase was detected by incubation with 100 µl/well of 1 mg/ml o-phenylenediamine and 1 µl/ml H_2O_2 in 0.1 M KH_2PO_4 buffer, pH 6.0. The reaction was stopped after 5 min by adding 100 µl/well of 1.3 N H_2SO_4 and the reaction product was quantified by measuring absorbance at 492 nm (reference wave length 405 nm) using an automated ELISA reader.

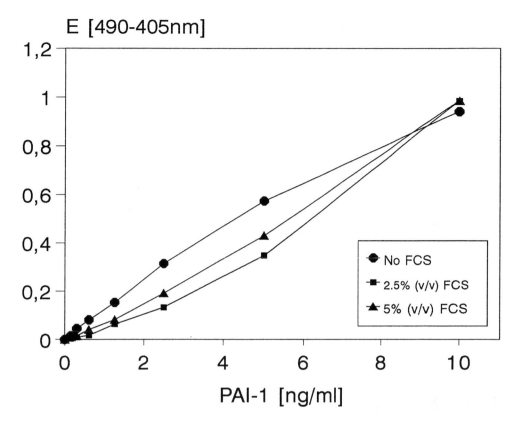

E [490-405nm]

PAI-1 [ng/ml]

Fig. 2. Standard curves for the PAI-1-specific ELISA in the absence or presence of FCS: No FCS (circles), 2.5% (v/v) FCS (squares) and 5% (v/v) FCS (triangles). The coefficients of correlation (r) for the different curves were as follows: No FCS r = 0.999, 2.5% FCS r = 0.988, 5% FCS r = 0.997.

Results

Selection of monoclonal antibodies for the PAI-1-specific ELISA

We aimed at establishing an immunoassay that detects PAI-1 in conditioned media, irrespective of whether the inhibitor was free or complexed with PAs or vitronectin. To select for the appropriate antibodies, several PAI-1-specific mAbs raised in our laboratory were tested for their reactivity with standard preparations of natural PAI-1, PAI-1/uPA complexes and PAI-1/tPA complexes by ELISA. Since PAI-1 can also form complexes with vitronectin [10] PAI-1/vitronectin complexes were also included in the evaluation.

MAb HD-PAI-1 14.1 was chosen as detecting antibody in the PAI-1-specific ELISA: the use of this antibody resulted in the sensitive detection of free PAI-1, PAI-1/PA complexes, and PAI-1/vitronectin complexes (summarized in Table 2 and Fig. 1). There was no reactivity with free PAs or PAI-2 (data not shown).

ELISA for PAI-1

Increasing concentrations of PAI-1 (Fig. 2) were used to assess the detection of the inhibitors in the presence of FCS. PAI-1 was diluted in the absence (Fig. 2: circles) or presence of 2.5% (Fig. 2: squares) or 5% (Fig. 2: triangles) FCS. The linear dose-response relationship between the concentration of PAIs and the extinction values, as well as the absolute extinction values, were not affected by the presence of FCS.

An extinction value of three standard deviations above the mean calculated from nine replicate negative controls (= extinction values obtained in the absence of PAI-1) was arbitrarily defined as the positive/negative cut-off point ("detection limit"). The detection limit for the PAI-1-specific ELISA was ≈ 0.15 ng/ml (Table 2).

The intra-assay variation of the ELISA was explored according to a previously described method for the direct estimation of imprecision [11]. At least 90 different concentrations of each antigen were measured in double determinations and the corresponding coefficients of variations (C.V.) were calculated. For the PAI-1-specific ELISA concentrations ranging between 20 and 35 ng/ml were tested (Fig. 3). An acceptable degree of precision was arbitrarily defined as a C.V. below 10%: values

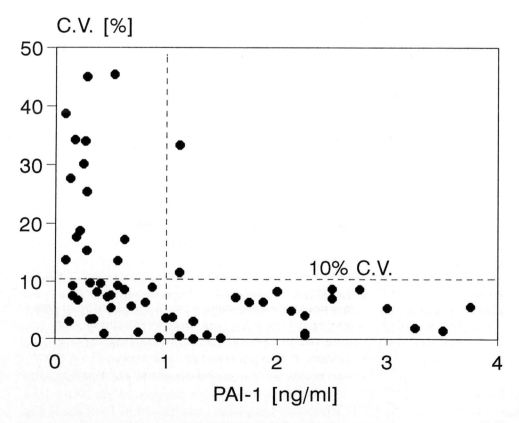

Fig. 3. Precision profiles of the PAI-1-specific ELISA determined according to Sadler et al. [11]. Sufficient precision is obtained with coefficients of variation (C.V.) below 10%.

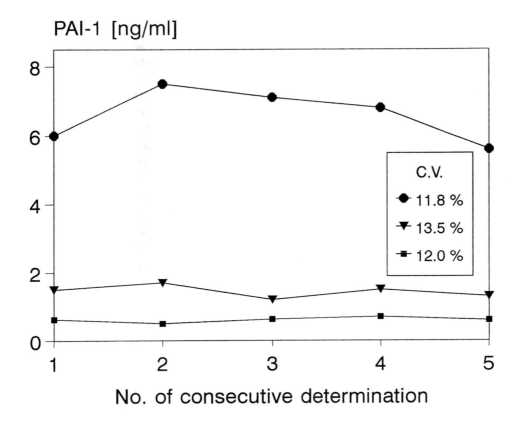

Fig. 4. Inter-assay-variability of the PAI-1-specific ELISA. High (7 ng/ml), intermediate (1.8 ng/ml) and low (0.5 ng/ml) concentrations of PAI-1 were measured at 5 consecutive days in double determinations. The inset shows the corresponding coefficients of variation (C.V.).

below the hatched lines in Fig. 3. The concentration value where the C.V. was below 10% was 1 ng of PAI-1/ml (Table 2).

The interassay variation was determined by testing a low (0.5 ng of PAI-1/ml), intermediate (1.8 ng of PAI-1/ml) and high (7 ng of PAI-1/ml) concentration of PAI-1 (Fig. 4) at 5 consecutive days. The C.V. ranged between 11.8% and 13.5%.

Detection of PAI-1 in conditioned media of the cell line HT-1080

PAI-1 was determined in the serum-free culture supernatants of the cell line HT-1080 cells, that is known to produce PAI-1 [8]. A constitutive release of PAI-1 was observed in HT-1080 cells: at a seeding density of 1×10^5 cells/ml a time-dependent increase of PAI-1 antigen in the conditioned medium was observed (Fig. 5). After 96 h of incubation 222.5 (± 15.5) ng/ml of PAI-1 were present in the conditioned medium.

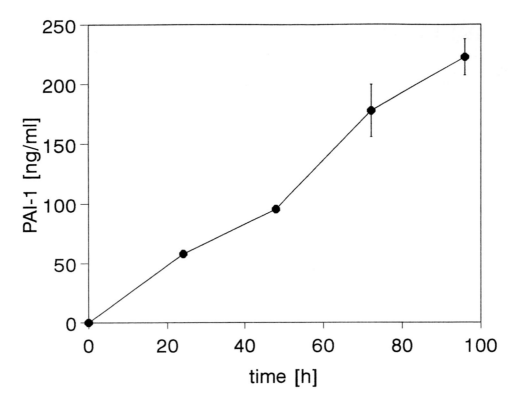

Fig. 5. PAI-1-antigen in conditioned media of HT-1080 cells. Serum-free supernatants were harvested after different intervals of time (abscissa).

Discussion

Here we have described an ELISA for the quantification of PAI-1. The immuno-logical reagents used to establish the assay were selected to allow the detection of PAI-1 even when the inhibitor was complexed with uPA, tPA or vitronectin. Usefulness of the assay was demonstrated by testing conditioned media derived from the PAI-1 producing cell line HT-1080.

The finding that PAI-1 detection was not influenced by complex formation with uPA, tPA or vitronectin (Fig. 1), indicates that the antigenic epitope recognized by mAb HD-PAI-1 14.1 is equally accessible in free as well as in complexed PAI-1.

The findings with HT-1080-conditioned media exemplify the usefulness of the enzyme immunoassay systems for the detection of PAI-1-antigen in cell culture supernatants. It is to be expected that the application of the assays will add to the analysis of PAI-1 producing cultured human cells. Due to its high level of sensitivity, which is comparable to previously published assay systems [12] and a high degree of precision even at low antigen concentrations (Fig. 3), the assay can be used to study the regulation of PAI-1 release by biological response modifiers, such as cytokines or pharmacological drugs.

Whether or not the assay can be used to quantify PAI-1 in human body fluids, the matrix of which is even more complex and variable than serum-containing cell culture media, has to await future studies. It will be of particular interest to assess the usefulness of the assay system for testing of tumor extracts, in which PAI-1 is suspected to be a marker for poor prognosis.

Acknowledgements

The authors are indebted to Uta Schirmer, Silke Jobstmann and Florentin Fedel for expert technical assistance. MDK is indebted to Dr Klaus Rother for continuous and generous support. The work has financially been supported by the Deutsche Forschungsgemeinschaft (Kr 931/2-2; Kr 931/3-1) and the Bundesministerium für Forschung und Technologie (01KG8808/4).

References

1. Saksela O. Biochim Biophys Acta 1985;823:35—65.
2. Danø K, Andreasen PA, Grondahl-Hansen J, Kristensen P, Nielsen LS, Skriver L. Adv Cancer Res 1985;44:139—266.
3. Vassalli J-D, Sappino A-P, Belin D. J Clin Invest 1991;88:1067—1072.
4. Scully MF. Br J Haematol 1991;79:537—543.
5. Andreasen PA, Georg B, Lund LR, Riccio A, Stacey SN. Mol Cell Endocrinol 1990;68:1—19.
6. Justus C, Müller S, Kramer MD. Br J Dermatol 1987;117:687—694.
7. Kramer MD, Vettel U, Schmitt M, Reinartz J, Brunner G, Meissauer A. Fibrinolysis 1992;6:103—111.
8. Lindahl T, Wiman B. Biochim Biophys Acta 1989;994:253—257.
9. Buessecker F, Reinartz J, Schirmer U, Kramer MD. J Immunol Meth 1993;162:193—200.
10. Keijer J, Linders M, Wegman JJ, Ehrlich HJ, Mertens K, Pannekoek H. Blood 1991;78(5): 1254—1261.
11. Sadler WA, Smith MH, Legge HM. Clin Chem 1988;34:1058—1061.
12. Declerck PJ, Alessi MC, Verstreken M, Kruithof EKO, Juhan-Vague I, Collen D. Blood 1988;71(1): 220—225.

©1994 Elsevier Science B.V. All rights reserved
Prospects in diagnosis and treatment of breast cancer
M. Schmitt et al., editors

Association of PAI-1 with metastasis-free survival in breast cancer: comparison with ER, PgR, PS2, cathepsin D and uPA

John A. Foekens[1], Manfred Schmitt[3], Harry A. Peters[1], Maxime P. Look[1], Wim L.J. van Putten[2], Michael D. Kramer[4], Fritz Jänicke[3] and Jan G.M. Klijn[1]

[1]Division of Endocrine Oncology, Department of Medical Oncology; [2]Department of Statistics, Dr Daniel den Hoed Cancer Center, P.O. Box 5201, 3008 AE Rotterdam, Netherlands; [3]Frauenklinik der Technischen Universität München; Klinikum rechts der Isar, 81675 München, Germany; and [4]Institute for Immunology, Heidelberg University, D-6900 Heidelberg, Germany

Abstract. Urokinase plasminogen activator (uPA), its inhibitor plasminogen activator inhibitor-1 (PAI-1), cathepsin D, estrogen receptor (ER) and its regulated proteins progesterone receptor (PgR) and PS2 have been reported to be associated with prognosis in primary breast cancer. The levels of all six parameters in cytosols of 587 primary breast tumor have been correlated with the length of distant metastasis-free survival (MFS). In univariate analysis, PAI-1, uPA and cathepsin D were all positively related with the rate of metastasis, whereas ER, PgR and PS2 were weakly associated with a more favorable prognosis. In Cox multivariate regression analysis, also including tumor size, nodal status, age and menopausal status, PAI-1 and uPA were the only cytosolic factors which significantly contributed to the model with relative hazard rates of 1.79 ($p < 0.0001$) and 1.41 ($p < 0.05$), respectively. It is concluded that the levels PAI-1 and uPA measured in routinely prepared cytosols are important parameters to predict the metastatic potential in patients with primary breast cancer.

Introduction

Mortality in breast cancer is related to the capacity of tumor cells to invade and to metastasize. Approximately 50% of patients with breast cancer will eventually die of metastatic disease. Adjuvant therapy cures an additional 4 to 10% of patients with node-negative and node-positive primary breast cancer, respectively [1]. However, if adjuvant treatment is given to all patients regardless of their lymph-node status, the majority of the patients will be overtreated. This is especially true for node-negative patients of whom only approximately 30% will experience a recurrence after locoregional treatment alone. In view of treatment burden of the patient, and efficacy/cost effectiveness, identification of patients at high risk for relapse is important [2–5].

Degradation of the extracellular matrix is facilitated by various proteases which allow the tumor cells to invade the surrounding tissue and to metastasize. Among these proteases are cathepsins and plasminogen activators [6,7]. Cathepsin D is a lysosomal acidic protease [8] being associated with an increased rate of relapse, particularly in node-negative patients [9–11]. The urokinase-type plasminogen activator (uPA) is a serine protease which catalyses the conversion of plasminogen into the active enzyme plasmin. Plasmin can activate type-IV collagenase which then degrades collagen and proteins of the basement membranes [7,12]. Recent studies have shown that high levels of uPA in primary breast tumors are associated with a poor prognosis for both node-negative and node-positive patients [13–20].

Breast tumors also contain plasminogen-activator inhibitor-1 (PAI-1) [16,19–24], a glycoprotein belonging to the family of the serine protease inhibitors, and which blocks the activity of uPA through formation of a covalent inhibitor-enzyme complex [25]. Therefore, it was an unexpected finding that high levels of PAI-1 in primary breast tumors were not associated with a better prognosis, but on the contrary were related with an increased relapse rate [16,19,20].

Recently we reported on the positive associations of the proteases uPA [17] and cathepsin D [26] with disease-free and overall survival rates in a series of approximately 700 patients. To evaluate the role of PAI-1, together with uPA, cathepsin D, ER and its regulated proteins PgR and PS2, we have measured PAI-1 in cytosols routinely prepared for steroid-hormone receptor determination. Here we report on the association between PAI-1 and the length of distant metastasis-free survival (MFS) in univariate and multivariate analyses of 587 patients, from whom the data on all six cytosolic parameters were available.

Patients and Methods

Patients

This study was performed on a group of 587 Dutch patients with operable primary breast cancer (between 1978 and 1987) and without signs of distant metastasis at surgery. Three hundred and thirty-two patients underwent modified mastectomy and 255 patients breast conserving lumpectomy. Two hundred and ten patients developed distant metastasis and 184 patients died during follow-up. Distant metastases included skin lesions not considered local, supraclavicular-, cervical-, and contralateral axillary- or parasternal lymph nodes, skeletal, and visceral spread.

Median follow-up time of patients still alive was 51 months. Adjuvant therapy was given exclusively to node-positive patients, i.e., to 97 of 108 node-positive premenopausal patients, and to 32 of 223 node-positive postmenopausal patients. None of the node-negative patients (n = 251) received adjuvant treatment. Of five patients the nodal status was unknown. Menopausal status was defined according to the criteria used by the European Organization for Research and Treatment of Cancer (EORTC) [27].

Tumors

Tumor specimens were drawn from a pool of frozen specimens (stored in liquid nitrogen) originally submitted to our laboratory for steroid hormone receptor analysis. Processing and pathological examination of the tumors were performed as described previously by Foekens et al. [28].

Assays

All assays were performed on cytosols routinely prepared for steroid receptor analysis as recommended by the EORTC [29]. Estimation of ER and PgR by radioligand binding assays, of PS2 and cathepsin D by radiometric immunoassays and of uPA by enzyme-linked immunosorbent assay (ELISA) were as described by Foekens et al. [17,26,30]. PAI-1 was determined with an ELISA, using polyclonal antibody to human PAI-1 (#395-G, American Diagnostica, Greenwich, CT) as catching antibody and culture supernatant of monoclonal antibody HD-PAI-1 14.1, as detecting antibody. Peroxidase conjugated goat anti mouse polyclonal antibody and the 1,2-phenylenediamine reaction was used to determine the amount of PAI-1 with human PAI-1 (#1090, American Diagnostica) as standard. The anti-PAI-1 monoclonal antibody HD-PAI-1 14.1 recognizes both active and inactive forms of PAI-1 and in addition it recognizes PAI-1 complexed with urokinase, tissue type plasminogen activators, or with vitronectin.

Statistics

MFS probabilities were calculated by the actuarial method of Kaplan and Meier [31]. The Cox proportional hazard model was applied for both univariate and multivariate analyses, with the associated likelihood ratio tests or t-tests used for tests of difference or trend. The associations between the cytosolic factors were described by Spearman rank correlation coefficients.

Results

Associations between PAI-1 and other variables

The median values of PAI-1 and cathepsin D were 15.2 ng/mg cytosolic protein and 45.3 pmol/mg protein, respectively. These values were arbitrarily used as cut-off points. For uPA and PS2 the used cut-off points were 1.15 and 2.0 ng/mg protein, respectively, as established before in analyses for disease-free survival [17,26]. The levels of PAI-1 were positively correlated with those of uPA (Spearman $R_s = 0.60$, $p < 0.001$) and cathepsin D ($R_s = 0.34$, $p < 0.001$), and negatively with those of ER ($R_s = -0.09$, $p < 0.05$), PgR ($R_s = -0.15$, $p < 0.01$) and PS2 ($R_s = -0.14$, $p < 0.01$). Also the levels of uPA and cathepsin D were found to be positively correlated ($R_s = 0.40$, $p < 0.01$). PAI-1 was not significantly correlated with tumor size, lymph-node status or menopausal status and age.

Univariate analysis for 7-year MFS

Tumor size and nodal status were both strongly associated with MFS ($p < 0.0001$). The relationship of age and menopausal status with MFS did not reach statistical significance. The association between PS2 positivity and a favorable MFS was of a similar magnitude (relative hazard rate, RHR = 0.79) as that of the combined ER/PgR status (RHR = 0.72, p = 0.02), but was statistically not significant (p = 0.09). High levels of uPA and cathepsin D were associated with a shorter MFS, both when analyzed dichotomized (for both, p < 0.0001) and continuous as logarithmically transformed variable (p = 0.001 and p = 0.003, respectively). The association of PAI-1 with the length of MFS is visualized in Fig. 1.

Tumors containing high levels of PAI-1 showed a significantly shorter MFS (p < 0.0001) as compared with tumors containing PAI-1 levels below the median value of 15.2 ng/mg protein. Also in Cox univariate regression analysis using logarithmically transformed PAI-1 values, PAI-1 was found to be significantly associated with the rate of occurrence of distant metastasis (p < 0.0001).

Multivariate analysis for MFS

To assess the prognostic value of PAI-1 with respect to other factors, PAI-1 status was combined with clinical parameters (age, menopausal status, tumor size and nodal status) and the other cytosolic markers (ER, PgR, PS2, cathepsin D and uPA) in Cox multivariate analysis for 7-year MFS. After correction for the clinical factors, neither ER, PgR nor PS2 contributed to the model. In further Cox analyses ER, PgR and PS2 were combined and added as a dichotomized variable (all three positive vs. all other phenotypes). The results of the multivariate analysis for 7-year MFS after inclusion of all cytosolic factors together as dichotomized variables are listed in Table 1.

Table 1. Multivariate regression analysis for 7-year MFS

Factor	p value	Relative failure rate (95% confidence limits)
Age and menopausal status	0.18	
Age postmenopausal (decades)		0.75 (0.53–1.07)
Age premenopausal (decades)		0.85 (0.68–1.05)
Menopausal status (post vs. pre)		1.79 (1.01–3.16)
Tumor size: 2–5 cm vs. ≤2 cm	0.07	1.37 (0.98–1.92)
>5 cm vs. ≤2 cm	<0.005	1.94 (1.25–3.02)
Nodal status: 1–3 vs. 0	<0.005	1.74 (1.20–2.54)
>3 vs. 0	<0.0001	2.82 (1.98–4.02)
ER/PgR/PS2 (+/+/+ vs. others)	0.10	0.78 (0.58–1.05)
Cathepsin D (>median vs. ≤median)	0.23	1.20 (0.89–1.64)
uPA (positive vs. negative)	<0.05	1.41 (1.03–1.93)
PAI-1 (>median vs. ≤median)	<0.0001	1.79 (1.29–2.49)

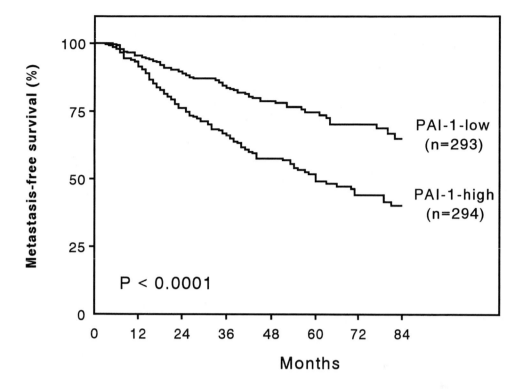

Fig. 1. Actuarial MFS curves stratified by PAI-1 status. PAI-1-low, < 15.2 ng/mg protein; PAI-1-high, ≥ 15.2 ng/mg protein.

Tumor size and nodal status strongly added to the model, whereas the contribution of the combined age and menopausal status was only marginal. Of the cytosolic markers only uPA (p < 0.05) and PAI-1 (p < 0.0001) remained in the model to predict the rate of metastasis.

After inclusion of cathepsin D, uPA and PAI-1, as continuous variables in the Cox model, PAI-1 was the only cytosolic factor remaining significant (p = 0.001), together with tumor size and nodal status.

Combination of cathepsin D, uPA and PAI-1 in analysis for MFS

For visualization of the combined effects of cathepsin D, uPA and PAI-1 on the length of MFS, 7-year Kaplan-Meier curves were constructed after dividing the patients in four different groups based on the tumor status of cathepsin D, uPA and PAI-1. Fig. 2 shows that there was a 42% difference in the actuarial rate of metastasis in favor of patients with tumors negative for all three parameters (curve a) as compared with those with tumors containing high levels of cathepsin D, uPA and PAI-1 (curve d).

As can be concluded from the multivariate analysis shown in Table 1 and from

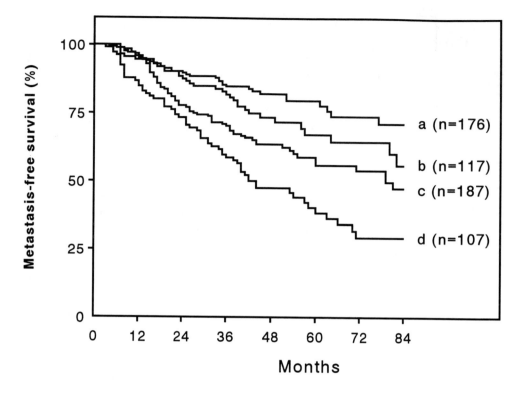

Fig. 2. Actuarial MFS curves stratified by PAI-1, uPA and cathepsin D status. a) Tumors with low levels of PAI-1, uPA and cathepsin D; b) Tumors with low levels of PAI-1, but high levels of uPA or cathepsin D, or both; c) Tumors with high levels of PAI-1, but low levels of uPA or cathepsin D, or both; d) Tumors with high levels of PAI-1, uPA and cathepsin D.

curve b representing patients with tumors having low levels of PAI-1, but high levels of cathepsin D or uPA or both, and curve c representing patients with tumors containing high levels of PAI-1, but low levels of cathepsin D, uPA or both (Fig. 2), the relationship of PAI-1 status to the rate of metastasis is the strongest. By additional stratification of ER/PgR/PS2-positivity an approximately 10% difference in probability of the occurrence of metastasis was observed (curves not shown).

Discussion

Proteases are of special importance in cancer research as they play a key role in the processes of invasion and metastatis. Cathepsin D and uPA are among the potentially important proteases. There have been several reports concerning the cathepsin D level in cytosols from primary breast tumor biopsies and its relationship with poor prognosis [6,9–11]. A positive relationship between the levels of uPA measured in primary breast tumor extracts or cytosols and poor prognosis has also recently been reported in studies involving 52 to 319 patients [13–16,19,20]. We have recently confirmed in a large series of approximately 700 Dutch patients with breast cancer that high levels of the proteases cathepsin D [26] and uPA [17] in cytosolic fractions of the primary tumor biopsies were associated with a poor prognosis. As the plasminogen-activator inhibitor-1 PAI-1 is known to block the activity of uPA [25], we have investigated in 587 of these patients the combined prognostic value of PAI-1 and other cell biological parameters reported to be positively (cathepsin D, uPA) or negatively (ER, PgR, PS2) associated with the rate of relapse. In the present study we have restricted our analyses on the relationships of the cytosolic parameters with the rate of the occurrence of distant MFS.

We observed a strong positive correlation between the levels of PAI-1 and uPA, an observation also made by other investigators [16,19,20,22–24]. Moreover, we observed a significant positive correlation between the levels of uPA and cathepsin D, which is in agreement with other authors also assaying routinely prepared cytosols [18,22,32], but it is in contrast to the results obtained with triton-extracts, showing no correlation [19]. The observed positive correlation between the levels of PAI-1 and cathepsin D in the present study is in agreement with the results reported by Foucré et al. [22].

In univariate regression analysis PAI-1, uPA and cathepsin D were all positively associated with the rate of occurrence of distant metastasis, both when analyzed as continuous and as dichotomized variables. For dichotomy, in these analysis an optimized cut-off point was used for uPA [17], and the median PAI-1 and cathepsin D values were arbitrarily chosen as cut-off points to dicriminate between tumors containing high and low levels. ER, PgR and PS2 were all negatively associated with MFS, but the associations were much weaker than previously obtained in analyses for disease-free and overall survival [26].

Comparing various cytosolic parameters for their relationship with the rate of distant metastasis with Cox multivariate analysis, also including nodal status, tumor size, age and menopausal status, PAI-1 and uPA were the only independent cytosolic factors contributing to the model. PAI-1 was the strongest factor ($p < 0.0001$), followed by uPA ($p < 0.05$). In the study of Jänicke et al. [19], uPA was found to be a stronger discriminator than PAI-1 in the multivariate analysis for relapse-free survival. Measurement of uPA and PAI-1 in tumor tissue extracts prepared in different ways, and the different endpoints used in both analyses, i.e., MFS vs. relapse-free survival, may have caused the observed disparity.

It was surprising that in the present and other smaller studies [16,19,20], high PAI-1 levels were found to be associated with a poor prognosis as PAI-1 is known to inhibit both free and receptor-bound uPA [25,33]. Different possible mechanisms to explain this phenomenon have been proposed. It has been argued that PAI-1 may be produced as a mechanism of self protection against the uPA directed tissue destruction [23,34], or could be a biochemical measure of the degree of neovascularization [20,34]. Moreover, the excess release of PAI-1 could be important for the reimplantation of circulating tumor cells at the metastatic site [19].

In the present study of six cytosolic cell biological parameters, we have shown that PAI-1 is the strongest predictor of time to the occurrence of distant metastases. It is concluded that knowledge of the PAI-1 status of the primary breast tumor may be useful in the refinement of adjuvant treatment strategies and for development of new treatment modalities as shown for uPA [35,36].

Acknowledgements

We are indebted to Mr H. Portengen, Ms Erica Binnendijk-Noordegraaf, Ms Marly Stuurman-Smeets, Ms Yolanda van Kooij, and Mr Piet van Assendelft for excellent technical assistance, to Helen de Koning, MD, M.J. Hooning, MD, Marijke Bontenbal, MD, and Jana Alexieva-Figusch, MD, PhD, for their assistance in the collection of the clinical data, and to Bruno Thirion, MD, PhD, CIS bio international, for generously providing assay kits for PS2 and cathepsin D. Work done at the Dr Daniel den Hoed Cancer Center in Rotterdam was supported by grant DDHK 92-04 from the Dutch Cancer Society (NKB), and work done at the Technische Universität München was supported by grants GR280/4-1 of the Klinische Forschungsgruppe der Deutschen Forschungsgemeinschaft and by the Wilhelm-Sander-Stiftung.

References

1. Early Breast Cancer Trialists' Collaborative Group: Systemic treatment of early breast cancer by hormonal, cytotoxic or immune therapy. Lancet 1992;339:1—15,71—85.
2. Hillner BE, Smith TJ. N Engl J Med 1991;324:160—168.
3. McGuire WL, Clark GM. N Engl J Med 1992;326:1756—1761.
4. McGuire WL, Tandon AK, Allred DC, Chamness GC, Ravdin PM, Clark GM. Cancer 1992;70: 1775—1781.
5. Gasparini G, Pozza F, Harris AL. J Natl Cancer Inst 1993;85:1206—1219.
6. Rochefort H, Capony F, Garcia M. Cancer Metastasis Rev 1990;9:321—331.
7. Danø K, Andreasen PA, Grøndahl-Hansen J, Kristensen PI, Nielsen LS, Skriver L. Adv Cancer Res 1985;44:139—266.
8. Rochefort H, Capony F, Garcia M. J Cell Biochem 1987;35:17—29.
9. Thorpe S, Rochefort H, Garcia M, Freiss G, Christensen I, Khalaf S, Paolucci F, Pau B, Rasmussen B, Rose C. Cancer Res 1989;49:6008—6014.
10. Spyratos F, Maudelonde T, Brouillet JP, Brunet M, Defrenne A, Andrieu C, Hacène K, Desplaces A, Rouesse J, Rochefort H. Lancet 1989;ii:1115—1118.
11. Tandon AK, Clark GM, Chamness GC, Chirgwin J, McGuire WL. N Engl J Med 1990;322:297—302.
12. Liotta LA, Goldfarb RH, Brundage R, Siegal GP, Terranova V, Garbisa S. Cancer Res 1981;41: 4629—4636.
13. Duffy MJ, O'Grady P, Devaney D, O'Siorain L, Fennelly JJ, Lijnen HJ. Cancer 1988;62:531—533.
14. Duffy MJ, Reilly D, O'Sullivan C, O'Higgins N, Fennelly JL, Andreasen P. Cancer Res 1990;50: 6827—6829.
15. Jänicke F, Schmitt M, Hafter R, Hollrieder A, Babic R, Ulm K, Gössner W, Graeff H. Fibrinolysis 1990;4:69—78.
16. Jänicke F, Schmitt M, Graeff H. Sem Thromb Hemostas 1991;17:303—312.
17. Foekens JA, Schmitt M, van Putten WLJ, Peters HA, Bontenbal M, Jänicke F, Klijn JGM. Cancer Res 1992;52:6101—6105.

18. Spyratos F, Martin P-M, Hacène K, Romain S, Andrieu C, Ferrero-Poüs M, Deytieux S, Le Doussal V, Tubiana-Hulin M, Brunet M. J Natl Cancer Inst 1992;84:1266–1272.
19. Jänicke F, Schmitt M, Pache L, Ulm K, Harbeck N, Graeff H. Breast Cancer Res Treat 1993;24: 195–208.
20. Grøndahl-Hansen J, Christensen IB, Rosenquist C, Brünner N, Mouridsen HT, Danø K, Blichert-Toft M. Cancer Res 1993;53:2513–2521.
21. Sumiyoshi K, Baba S, Sakaguchi S, Urano T, Takada Y, Takada A. Thromb Res 1992;63:59–71.
22. Foucré D, Bouchet C, Hacène K, Pourreau-Schneider N, Gentile A, Martin PM, Desplaces A, Oglobine J. Br J Cancer 1991;64:926–932.
23. Reilly D, Christensen L, Duch M, Nolan N, Duffy MJ, Andreasen PA. Int J Cancer 1992;50: 208–214.
24. Sumiyoshi K, Serizawa K, Urano T, Takada Y, Takada A, Baba S. Int J Cancer 1992;50:345–348.
25. Hekman CM, Loskutoff DJ. Semin Thromb Hemost 1987;13:514–527.
26. Foekens JA, van Putten WLJ, Portengen H, de Koning YWCM, Thirion B, Alexieva-Figusch J, Klijn JGM. J Clin Oncol 1993;11:899–908.
27. EORTC Breast Cancer Cooperative Group. Manual for clinical research in breast cancer. Van der Poorten, Belgium, 1991.
28. Foekens JA, Portengen H, van Putten WLJ, Trapman AMAC, Reubi J-C, Alexieva-Figusch J, Klijn JGM. Cancer Res 1989;49:7002–7009.
29. EORTC Breast Cancer Cooperative Group. Eur J Cancer 1980;16:1513–1515.
30. Foekens JA, Portengen H, van Putten WLJ, Peters HA, Krijnen HLJM, Alexieva-Figusch J, Klijn JGM. Cancer Res 1989;49:5823–5828.
31. Kaplan EL, Meier P. J Am Stat Assoc 1958;53:457–48
32. Duffy MJ, Brouillet J-P, McDermott E, O'Higgins NO, Fennelly JJ, Maudelonde T, Rochefort H. Clin Chem 1991;37:101–104.
33. Ellis V, Wun T-C, Behrendt N, Rønne E, Danø K. J Biol Chem 1990;265:9904–9908.
34. Pyke C, Kristensen P, Ralfkiær E, Eriksen J, Danø K. Cancer Res 1991;51:4067–4071.
35. Ossowski L, Russo-Payne H, Wilson EL. Cancer Res 1991;51:274–281.
36. Kobayashi H, Ohi H, Shinohara M, Fujii T, Terao T, Schmitt M, Goretzki L, Chucholowski N, Jänicke F, Graeff H. Br J Cancer 1993;67:537–544.

Urokinase (uPA) and PAI-1 as selection criteria for adjuvant chemotherapy in axillary node-negative breast cancer patients

F. Jänicke, Ch. Thomssen, L. Pache, M. Schmitt and H. Graeff

Frauenklinik der Technischen Universität München, Klinikum rechts der Isar, München, Germany

Abstract. The relative impact of prognostic factors on disease-free and overall survival reflects their respective role in tumor biology. Breast cancer can be taken as an example to demonstrate the clinical and tumor-biological relevance of tumor-associated proteases. Receptor-bound urokinase-type plasminogen activator (uPA) and its inhibitor PAI-1 seem to play a major role in the dissolution and formation of tumor stroma. These processes are prerequisites for invasion and metastasis. The evaluation of "classical" and new prognostic factors shows that the uPA and PAI-1 contents of breast cancer tissue are strong and independent prognostic factors. High-risk patients can even be identified within the classical risk groups defined by lymph node or hormone receptor status, thus allowing a more individualized delineation of those patients at low or high risk for relapse and death, irrespective of established risk factors.

Since high risk patients can be identified within the node-negative subgroup, the question arises whether these patients should be treated by adjuvant chemotherapy.

Whether high levels of uPA and PAI-1, which put patients into a high risk group, may affect response to adjuvant chemotherapy can only be answered by a clinical trial. Therefore we initiated a prospective randomized multicenter study in Germany, supported by the "Deutsche Forschungsgemeinschaft" (DFG), in which patients with high values of uPA and/or PAI-1 are randomized to six cycles CMF vs. observation. Patients with low content of both uPA and PAI-1 are distributed to observation only. Concomitantly other new tumor-biological factors like HER-2/neu, EGF-R, Ki-67 (MIB-I), p53, the cathepsins D, B and L and collagenases (type IV) will be determined in parallel to uPA and PAI-1 to compare their relative prognostic impact and to analyze their predictive value concerning chemosensitivity or -resistance in adjuvant treatment of node-negative patients.

In the western world one in eight women will experience breast cancer at some time in her life. Unfortunately, at the time of diagnosis the majority of patients has occult micrometastatic disease due to a marked capability of tumor cells for invasion and early hematogenic spread. Therefore every second patient with breast cancer will develop metastases months or years after first diagnosis and finally succumb to the disease. On the other hand, breast cancer is a heterogeneous disease showing great variability of biological and clinical behavior.

In an extensive meta-analysis it was clearly shown that adjuvant therapy improves the course of the disease in breast cancer, evidently even after prolonged time of observation. In this analysis it was calculated that adjuvant chemotherapy reduces relapse rate by 28% and adjuvant tamoxifen therapy by 25% [1]. In other words, on average one out of four recurrences will be prevented by applying adjuvant treatment. This risk reduction was observed independent of the stage of disease with the same percentage in node-negative and node-positive patients. As a consequence recommen-

NODE - NEGATIVE BREAST CANCER

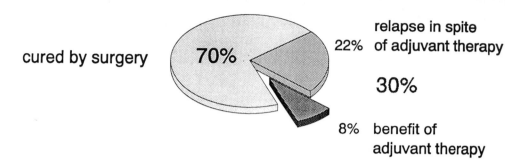

cured by surgery   70%

22%   relapse in spite of adjuvant therapy

30%

8%   benefit of adjuvant therapy

Fig. 1. Effect of unselected adjuvant therapy in node negative breast cancer patients. Provided a 30% relapse rate and a 26% reduction in relapses as calculated in the meta-analysis, absolute benefit for the individual patient is 8%. In 22% of the patients relapse will occur due to resistance to therapy.

dations were given to offer all patients some kind of adjuvant therapy regardless of the lymph node status. However, if benefit is calculated in terms of absolute improvement for the individual patient, benefit increases in parallel to the individual risk of the patient and to the sensitivity of the individual tumor towards chemo- or hormonal therapy.

In node-negative disease prognosis is generally assumed to be good, as about 70% of the patients are cured by locoregional surgery alone. In view of this, considerable controversy remains regarding adjuvant chemotherapy in women with early stages of the disease, especially in patients with tumor-free axillary lymph nodes. Assuming an average risk reduction of 26% in node-negative patients, an absolute improvement of only 8% will be achieved if all patients are given adjuvant treatment in an unselected manner (Fig. 1). Alternatively, high-risk patients should be selected for adjuvant chemotherapy by a combination of different prognostic factors [2]. A majority of women with node-negative disease would thereby be spared unnecessary and potentially toxic adjuvant therapy. Unfortunately up to now neither conventional nor new prognostic factors are of satisfactory value for clinical decision making in node-negative breast cancer.

Tumor size is an important prognostic indicator in axillary node-negative patients, as tumors <1 cm indicate an excellent prognosis [2]. On the other hand, tumor diameters >1 cm have only a minor effect on the 5-year survival rate, indicating that size may merely reflect the age of the tumor and to a lesser degree its aggressiveness [2–4].

The prognostic value of steroid hormone receptor status is considerably lower than that of lymph node status. With increasing time of follow-up the initially divergent relapse-free survival curves of positive and negative steroid hormone receptors begin to converge. Although highly statistically significant in node-negative disease, the difference between estrogen-receptor positive and estrogen-receptor negative tumors with regard to 5-year survival amounts to only about 8–9% [2]. Therefore hormone receptor status is not a sufficiently strong prognostic factor in node-negative breast cancer patients to warrant basing treatment decisions on it alone.

According to Clark new prognostic factors should meet strict requirements before clinical application to therapeutic decisions is justified. Firstly, a sensitive, specific

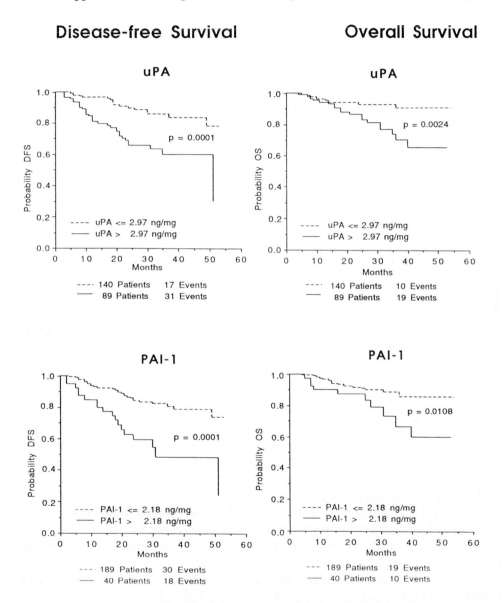

Fig. 2. Disease-free and overall survival as a function of uPA or PAI-1 in breast cancer patients (n = 229). Patients with high uPA or PAI-1 content had a significantly lower rate of disease-free and overall survival than patients with low content of uPA or PAI-1 (From [37], with permission from Kluwer Publishers).

Table 1. Prognostic impact of uPA (relative risk[a]) within subgroups of breast cancer patients

Subgroup	Number	Relative risk (95%) CI[b]	p value[c]
All patients	229	3.22 (1.8— 5.8)	0.0001
Node-negative	101	5.1 (1.4—18.9)	0.01
Node-positive	128	2.74 (1.4—11.3)	0.002
Hormone receptor-positive	170	3.72 (1.8— 7.9)	0.0002
Hormone receptor-negative	59	1.79 (0.7— 4.5)	0.213
Pre-/perimenopausal	90	3.46 (1.6— 7.6)	0.001
Postmenopausal	139	3.12 (1.3— 7.6)	0.008

[a]RR was calculated by Cox Regression Analysis; [b]CI = confidence interval; [c]cut-off for uPA: 2.97 ng/mg protein.

detection method for the factor in question has to be developed and measured in tumor tissue. This new factor must then be correlated with established prognostic factors and the follow-up of a patient collective. Statistical analyses should include univariate analysis (significance for relapse-free as well as overall survival), correlation to other prognostic factors and multivariate analysis for comparison with other factors (i.e., rank, relative risk, independence of the new factor). Validation of the results in different sets of patients, preferably by other researchers, is also highly recommended. Only if all of these requirements are fulfilled can one estimate whether the prognostic factor in question will be of clinical relevance in terms of quality and reliability of the prognosis.

The malignant potential of solid tumors consists of their capacity for fast proliferation on the one hand and for invasion and metastasis on the other. The proliferation rate in breast cancer can now be determined by cellular parameters such as S-phase fraction, thymidine labelling index (TLI) or Ki-67 (MIB-I) antigen. Steroid receptors (inverse correlation) and EGF-R show a high correlation with the above proliferation parameters; they can therefore be regarded as indirect indicators of proliferation.

Currently S-phase and ploidy measurements are predominantly performed by flow cytometry using fluorescent markers. A major disadvantage of flow cytometry is the fact that in aneuploid tumors (about two thirds of all patients) determination of S-phase becomes difficult in those cases where an overlap of histograms is present. However, in subgroups of patients, S-phase and ploidy analysis seem to be reliable indicators of cell kinetics and thus valuable prognostic markers for node-negative breast cancer patients [4,5].

The so-called "thymidine-labelling-index" (TLI) describes the uptake of thymidine by cells currently in the S-phase of cell division (phase of DNA synthesis). TLI was reported to be an important prognostic factor in studies with a long time follow-up of more than 10 years [6—8]. Broad clinical application of the method is hampered by the fact that fresh tumor tissue has to be incubated with a radioactive tracer molecule.

In 1983 Gerdes [9] developed Ki-67, a monoclonal antibody that recognizes a nucleus-associated antigen present in proliferating cells but absent in dormant cells.

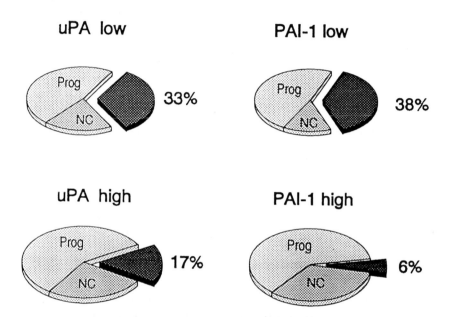

DISSEMINATED BREAST CANCER
(n = 40)

RESPONSE (CR+PR) TO ENDOCRINE THERAPY

uPA low

PAI-1 low

Prog
NC
33%

Prog
NC
38%

uPA high

PAI-1 high

Prog
NC
17%

Prog
NC
6%

Fig. 3. Response of disseminated breast cancer to endocrine palliative treatment in relation to uPA and PAI-1 values in the primary tumor. Cut-off levels for uPA were 3 ng/mg protein; for PAI-1 1.6 ng/mg protein. CR = complete remission; PR = partial remission; NC = no change; Prog = progression. Evidently, hormone dependency declines with increasing tumoral uPA and PAI-1.

The percentage of cells expressing the antigen can be determined immunohistochemically. There is a positive correlation to S-phase fraction, grading, the expression of the HER2/neu oncoprotein and an inverse correlation to steroid hormone receptor status. A significant impact of Ki-67 on relapse-free as well as overall survival has been reported by different investigators only for the entire group of breast cancer patients. However, the data presented so far for the subgroup of node-negative breast cancer patients have not yet been conclusively analyzed [10,11].

The epidermal growth factor receptor (EGF-R) can be detected in cell membranes of tumor cells by radioligand assay, enzyme-linked immunosorbent assay (ELISA) or immunohistochemistry. EGF-R is inversely correlated to the estrogen receptor status and has a prognostic impact for node-positive as well as node-negative breast cancer patients. Even though early conclusive reports on the prognostic impact of

Table 2. Multivariate and univariate analysis of disease-free survival in 101 axillary node-negative breast cancer patients [37]

Variable	Univariate p	Multivariate p	Relative risk (95% CI[a])
uPA (>2.96 vs. ≤2.96 ng/mg protein)	0.0098	0.0136	5.5 (1.2–26.2)
PAI-1 (>2.18 vs. ≤2.18 ng/mg protein)	0.0036	0.0189	4.9 (1.4–17.2)
Vascular invasion (present vs. absent)	0.0267	0.2954	/
Cathepsin D (>50 vs. ≤50 pmol/mg protein)	0.0771	0.4687	/
Estrogen receptor status (<20 vs. ≥20 fmol/mg protein)	0.0503	0.6386	/
Hormone receptor status (negative vs. positive)	0.0712	0.6583	/
Tumor size (>1.4 cm vs. ≤1.4 cm)	0.6218	–	/
Progesterone receptor status (<20 vs. ≥20 fmol/mg protein)	0.9336	–	/

[a]CI = Confidence interval.

EGF-R have been controversial (presumably due to methodical difficulties), the prognostic impact of EGF-R in breast cancer does seem to be emerging [12].

A large number of studies demonstrate that expression of HER-2/neu (erbB2) is correlated with unfavorable prognosis mainly in node-positive patients [13,14]. In multivariate analysis, HER-2/neu ranks second to lymph node status in regard to its prognostic impact [15]. In node-negative patients no significant prognostic information is gained by detecting either overexpression of the oncogen (mRNA) or the oncoprotein itself (immunohistochemistry) [16].

It is remarkable that most of the prognostic factors currently under investigation in breast cancer are related to the capacity of the tumor for proliferation. Factors capable to quantify invasive and metastatic capacity of malignant cells should be even more suitable for estimation of prognosis, as mortality in the disease is predominantly based on occurrence of distant metastasis. Evidence has accumulated that invasion and metastasis in solid tumors require the action of tumor-associated proteases which promote the dissolution of the tumor matrix and the basement membranes [38]. This enables the tumor cells to invade the surrounding normal tissue, escape from their local tissue environment, enter the circulation and form distant metastases. A number of tumor-associated proteases such as cathepsins (D, B and L), metalloproteases

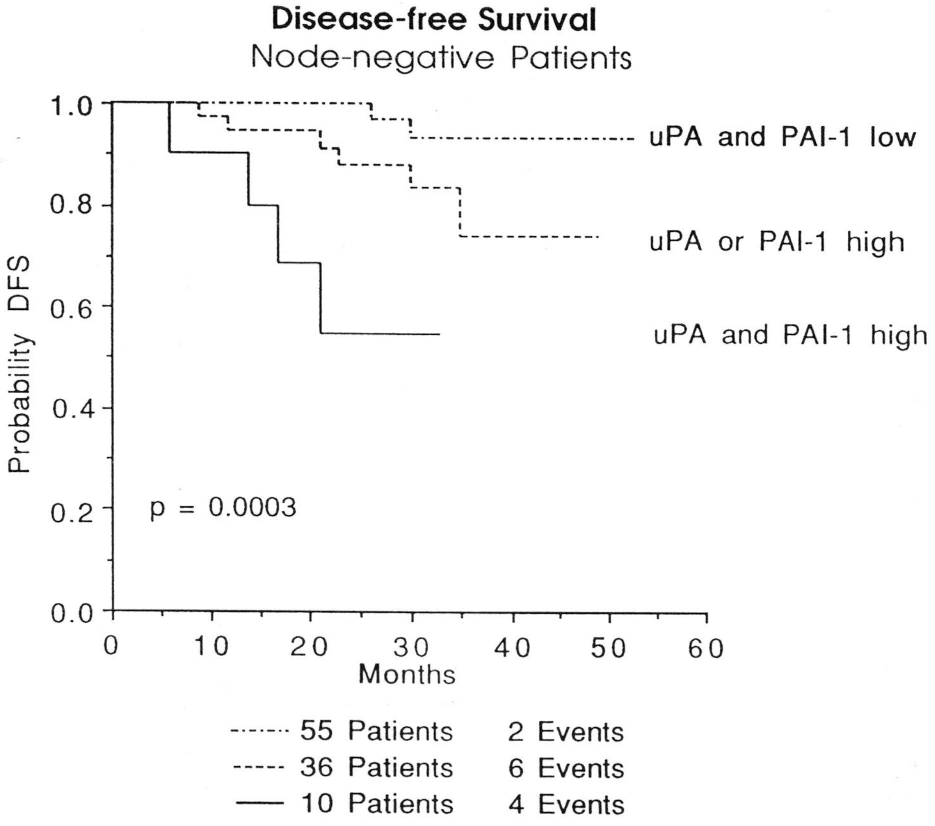

Fig. 4. Disease-free survival as a function of uPA plus PAI-1 in axillary node-negative breast cancer patients (n = 101). When both uPA and PAI-1 are introduced into the prognostic evaluation, three groups of patients can be defined with increasing risk of relapse. A group of patients with particularly good prognosis (comprising 55% of the node-negative patients) is thus identified by low uPA and low PAI-content (From [37], by permission of Kluwer Academic Publishers).

(collagenases and stromelysins) and the plasmin/plasminogen activating system (plasmin, uPA, tPA, PAI-1, PAI-2) are involved in these processes activating each other in form of a proteolytic cascade [38].

Receptor-bound urokinase-type plasminogen activator (uPA) appears to play a key role in these processes [17–21]. Tumor cells synthesize and secrete uPA as an inactive proenzyme (pro-uPA) which binds to specific receptors on the tumor cell surface [19–23]. After binding, pro-uPA is activated by plasmin, cathepsin B [24] or cathepsin L [25]. Receptor-bound, active uPA converts plasminogen to plasmin. Subsequently plasmin is also bound to a different receptor on the tumor cell surface [26]. Plasmin then degrades components of the tumor stroma (e.g., fibrin, fibronectin, proteoglycans, laminin) and may activate procollagenase type IV, which then degrades collagen type IV, a major part of the basement membrane [27]. Hence, uPA

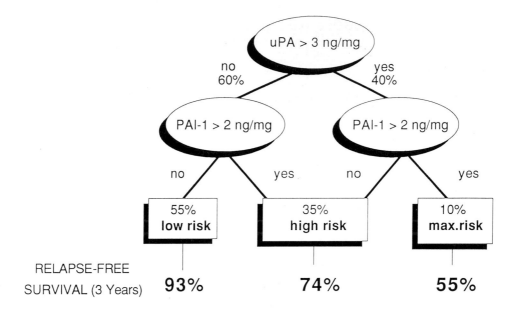

RELAPSE-FREE
SURVIVAL (3 Years)

Fig. 5. Patient selection by means of a sequential risk estimation (regression tree): uPA and PAI-1 as an example in axillary node-negative breast cancer. Division into subgroups with increasing risk is performed according to the respective impact of all independent factors determined by the Cox model, in this case beginning with uPA. The resulting subgroups are then further subdivided according to the second strongest factor (PAI-1).

will promote the dissolution of the tumor matrix and the basement membrane, which is a prerequisite for invasion and metastasis. However, tumors also exhibit the plasminogen activator inhibitor type 1 (PAI-1) which blocks the enzymatic activity of uPA. Thus the presence of both uPA and PAI-1 will modulate invasive and metastatic phenotype of cancer cells.

The strong and independent prognostic impact of uPA first observed by us [28–30] was later confirmed by others in independent studies [31-35]. Earlier observations applying activity measurements by Duffy et al. [36] already indicated a prognostic relevance of uPA. However, later studies by Jänicke et al. [28,29,37] and Duffy et al. [31,38] demonstrated that antigen measurements in breast cancer tissue extracts using ELISA techniques are far superior, and determined uPA antigen to be an independent prognostic factor.

High levels of the specific uPA-inhibitor PAI-1 were also found to be correlated with poor outcome in the disease. This initial observation by Jänicke et al. [30,37] was then also confirmed by others [35,39, Foekens (this volume)]. Patients with high levels of uPA or PAI-1 have a higher relapse rate and shorter survival time compared to those with low levels of one of these factors (Fig. 2). It seems somewhat contradictory that the uPA inhibitor PAI-1 is also an independent indicator of poor prognosis in breast cancer ranking close in order to uPA in the multivariate analysis. However, as PAI-1 has a protective role of cell-substratum vitronectin-dependent

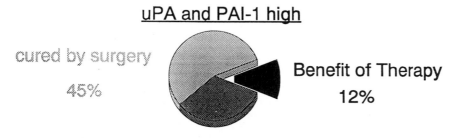

Fig. 6. Effectiveness of adjuvant therapy in risk groups of node negative breast cancer patients formed by uPA and PAI-1 determination in tumor tissue. Fifty-five percent of patients (with low uPA and PAI-1 levels) had less than a 10 % probability of relapse and thus are possible candidates for being spared the necessity of adjuvant therapy.

adhesions, the role of pericellular PAI-1 would also be to maintain contacts between migrating and invading cells and the matrix. The process of cell migration would thus need a balanced proteolysis in which a finely tuned protease-antiprotease equilibrium must be achieved. Moreover, excess release of PAI-1 into tumor tissue may also be of importance for the process of reimplantation of circulating tumor cells at distant loci after release from the primary site. When uPA activity is depressed by action of PAI-1, generation of metastases should be facilitated by formation of a new tumor stroma. These considerations lead one to suspect that if tumor cells of the primary tumor have a high capacity for both uPA and PAI-1 synthesis, they may also have a high capacity to metastasize.

In multivariate analysis of disease-free survival, uPA as an independent factor is shown to be as strong as the lymph node status [33,38], which has been claimed so far to be the strongest prognostic factor, followed by progesterone receptor status and PAI-1. Since uPA is an independent prognostic factor, high or low risk patients can be identified even within the classical risk groups defined by lymph node, hormone receptor or menopausal status. For example, within a group of 170 patients having hormone receptor positive tumors, who are supposed to have a good prognosis in general, about one third of them with a poor outlook can be identified by means of uPA measurement. These patients with high tumoral uPA levels are at a 3.7-fold risk to relapse compared to patients with low uPA values regardless of the fact that both

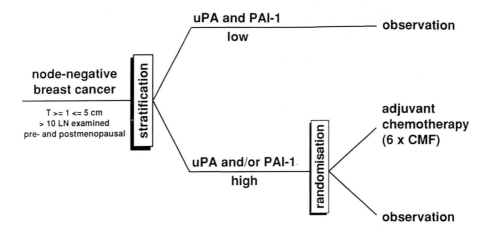

Fig. 7. Design of the prospective randomized German multicenter study of adjuvant chemotherapy in node-negative breast cancer patients using uPA and PAI-1 as selection criteria. Node-negative breast cancer patients with high values of uPA and/or PAI-1 are randomized to six cycles CMF vs. observation. Patients with low content of both uPA and PAI-1 are distributed to observation only.

have positive hormone receptors and receive adjuvant hormonal treatment (Table 1). This is also in line with the common observation that only about two-thirds of hormone receptor-positive patients respond to hormonal manipulation, and suggests that high uPA synthesis in breast cancer cells might be accompanied by expression of a "nonfunctional" hormone receptor and thereby loss of hormone dependency. This hypothesis is endorsed by our observation that patients having high uPA and/or PAI-1 content in their primary tumors, rarely respond to endocrine therapy when metastatic disease occurs (Fig. 3). Our findings have been confirmed by Klijn et al., who also found that high uPA levels indicated a poor response to endocrine therapy in more than 100 patients with metastatic breast cancer [40]. This is to demonstrate that new prognostic factors may not only be of use to give prognostic information but also to indicate sensitivity or resistance towards endocrine therapy or chemotherapy. In a similar manner HER-2/neu seems to indicate resistance to chemotherapy [41,42] and EGF-R resistance to tamoxifen [43], while high S-phase values may be indicators of chemosensitivity.

In node-negative patients the prognostic impact of uPA is closely followed by that of PAI-1, whereas hormone receptors, tumor size, vascular invasion and cathepsin D were of no significance in the multivariate setting (Table 2). The combination of the two independent variables uPA and PAI-1 should therefore result in an even more individualized delineation of those patients having a low or high risk of relapse. Indeed, the combination of both factors considerably enhances the prognostic value for node-negative patients (Fig. 4). The combination of uPA and PAI-1 allowed us to identify a group of 45% out of the node-negative breast cancer patients having an increased risk of relapse indicated by high uPA and/or high PAI-1 levels. In addition, this type of analysis allows the delineation of patients with a good prognosis (93%

disease-free survival at 36 months). Consequently, more than half of the patients had less than a 10% probability of relapse (Fig. 5). Provided the average risk reduction of 26% in node-negative patients [1], treatment of this low risk group would result in an absolute benefit of less than 2% for the individual patient. This calculated treatment effect seems to be of no clinical relevance. On the other hand, selective treatment of the high risk patients identified by high uPA/PAI-1 level could increase effectiveness of adjuvant treatment and help to avoid unnecessary and potentially toxic therapy in the majority of node-negative patients already cured by locoregional surgery alone (Fig. 6).

The question of whether high levels of uPA and PAI-1, which put patients into a high risk group, may affect response to adjuvant chemotherapy can only be answered by a clinical trial. In view of this, we initiated a prospective randomized multicenter study in Germany supported by the "Deutsche Forschungsgemeinschaft" (DFG), in which patients with high values of uPA and/or PAI-1 are randomized to six cycles CMF vs. observation. Patients with low content of both uPA and PAI-1 are distributed to observation only (Fig. 7). Concomitantly, other new factors like HER-2/neu, EGF-R, Ki-67 (MIB-I), p53, the cathepsins D, B and L and collagenases (type IV) will be determined in parallel to uPA and PAI-1 to compare their relative prognostic impact and to analyze their predictive value concerning chemosensitivity or resistance in adjuvant treatment of node-negative patients.

Acknowledgements

Supported by the Deutsche Forschungsgemeinschaft (Klinische Forschergruppe GR280/4-1) and the Wilhelm-Sander-Stiftung.

References

1. Early Breast Cancer Trialists' Collaborative Group (EBCTCG). Lancet 1992;339:1–15, 71–85.
2. McGuire WL, Tandon AK, Allred DC, Chamness GC, Clark GM. J Natl Cancer Inst 1990;82:1006–1015.
3. Tandon AT, Clark GM, Chamness GC, Chrigwin JM, McGuire WL. New Engl J Med 1990;322:297–302.
4. Sigurdsson H, Baldetorp B, Borg A, Dalberg M, Fernö M, Killander D, Olsson H. N Engl J Med 1990;322:1045–1053.
5. Clark GM, Dressler LG, Owens MA, Pounds G, Oldaker T, McGuire WL. N Engl J Med 1989;320:627–633.
6. Meyer JS. Natl Cancer Inst Monogr 1986;1:25.
7. Silvestrini R, Daidone MG, Valagussa P, Di Fronzo, G, Mezzanotte G, Bonadonna G. Eur J Cancer Clin Oncol 1989;25:1165.
8. Tubiana M, Pejovic MH, Koscielny S, Chavaudra N, Malaise E. Int J Cancer 1989;44:17.
9. Gerdes J, Schwab U, Lemke H, Stein H. Int J Cancer 1983;31:13.
10. Weikel W, Beck T, Mitze M, Knapstein PG. Breast Cancer Res Treat 1991;18:149.
11. Wintzer H, Zipfel I, Schulte-Monting J, Hellerich U, von Kleist S. Cancer 1991;67:421.
12. Klijn JGM, Berns PMJ, Schmitz PIM, Foekens J. Endocrine Rev 1992;13:3–17.

218

13. Anderson TJ. Human Pathol 1992;23:971—972.
14. Levine MN, Andrulis I. J Clin Oncol 1992;10:1034—1036.
15. Slamon DJ, Godolphin W, Jones LA, Holt JA, Wong SG, Keith DE, Levin WJ, Stuart SG, Udove J, Ullrich A, Press MF. Science 1989;244:707.
16. O'Reilly SM, Barnes DM, Camplejohn RS, Bartkova J, Gregory WM, Richards MA. Br J Cancer 1991;63:444.
17. Danø K, Andreasen PA, Grondahl-Hansen J, Kristensen P, Nielsen LS, Skriver, L. Adv Cancer Res 1985;44:139—266.
18. Markus G. Enzyme 1988;40:158—172.
19. Blasi F. Fibrinolysis 1988;2:73—84.
20. Mangel WF. Nature 1990;344:488—489.
21. Schlechte W, Murano G, Boyd D. Cancer Res 1989;49:6064—6069.
22. Stoppelli MP, Tacchetti C, Cubellis MV, Corti A, Hearing VJ, Cassani G, Appella E, Blasi F. Cell 1986;45:675—684.
23. Vassalli JD, Baccino D, Belin D. J Cell Biol 1985;100:86—92.
24. Kobayashi H, Schmitt M, Goretzki L, Chucholowski N, Calvete J, Kramer M, Günzler WA, Jänicke F, Graeff H. J Biol Chem 1990;266:5147—5152.
25. Goretzki L, Schmitt M, Mann KH, Calvete J, Chucholowski N, Kramer M, Günzler WA, Jänicke F, Graeff H. FEBS Lett 1992;297:112—118.
26. Miles LA, Plow EF. Fibrinolysis 1988;2:61—71.
27. Trygvasson K. In: Liotta LA (ed) Influence of Tumor Development on the Host. Dordrecht: Kluwer Academic Publishers, 1989;72—83.
28. Jänicke F, Schmitt M, Ulm K, Gössner W, Graeff H. Lancet 1989;8670:1049.
29. Jänicke F, Schmitt M, Hafter R, Hollrieder A, Babic R, Ulm K, Gössner W, Graeff H. Fibrinolysis 1990;4:69—78.
30. Jänicke F, Schmitt M, and Graeff H. Sem Thromb Hemostas 1991;17:303—312.
31. Duffy MJ, Reilley D, O'Sullivan C, O'Higgins N, Fennelly JJ, Andreasen P. Cancer Res 1990;50: 6827—6829.
32. Duffy MJ, Reilly D, O'Sullivan C, O'Higgins N, Fennelly JJ. Lancet 1990;335(8681):108.
33. Foekens JA, Schmitt M, van Putten WLJ, Peters HA, Brontenbal M, Jänicke F, Klijn JGM. Cancer Res 1992;52:6101—6105.
34. Spyratos F, Martin PM, Hacene K, Romain S, Andrieu C, Ferrero-Poüs M, Deytieux S, Le Doussal V, Tubiana-Hulin M, Brunet M. J Nat Cancer Inst 1992;84:1266—1272.
35. Grondahl-Hansen J, Christensen IJ, Rosenquist C, Brünner N, Mouridsen HT, Danø K, Blichert-Toft M. Cancer Res 1993;53:2513—2521.
36. Duffy M, O'Grady P, Devaney D, O'Siorain L, Fennelly JJ, Lijnen HJ. Cancer 1988;62:531—533.
37. Jänicke F, Schmitt M, Pache L, Ulm K, Harbeck N, Höfler H, Graeff H. Breast Cancer Res Treat 1993;24:195—208.
38. Schmitt M, Jänicke F, Graeff H. Fibrinolysis 1992;6(suppl 4):3—26.
39. Foekens JA, Schmitt M, van Putten WLJ, Peters HA, Jänicke F, Klijn JGM, J Clin Oncol 1994 (in press).
40. Klijn JG, Berns EM, Bontenbal M, Foekens J. Cancer Treat Rev 1993;19(suppl B4):5—63.
41. Allred DC, Clark GM, Tandon AK, Molina R, Tormey DC, Osborne CK, Gilchrist KW, Mansour EG, Abeloff M, Eudy L, McGuire WL. J Clin Oncol 1992;10(4):599—605.
42. Gusterson BA, Gelber RD, Goldhirsch A, Price KN, Säve-Söderborgh J, Anbazhagan R, Styles J, Rudenstam C-M, Golouh R, Reed R, Martinez-Tello F, Tiltman A, Torhorst J, Grigolato P, Bettelheim R, Neville AM, Bürki K, Castiglione M, Collins J, Lindtner J, Senn H-J. J Clin Oncol 1992;10(7):1049—1056.
43. Nicholson S, Sainsbury JRC, Halcrow P, Chambers P, Farndon JR, Harris AL. Lancet 1989;1:182.

Importance of menopausal status in breast cancer

Susan M. Thorpe[*]

Novo Nordisk, Novo Nordisk Park, 2760-DK Måløv, Denmark

Abstract. Menopausal status is generally not considered when evaluating significance of prognostic factors in breast cancer. The importance of menopausal status as a biologically relevant subgrouping in data analyses of breast cancer patients is discussed using examples from the Danish Breast Cancer Cooperative Group (DBCG) studies. Failure to consider menopausal status in data analyses can lead to erroneous conclusions.

Optimal implementation of prognostic factors for determining optimal treatment remains a primary concern of breast cancer physicians. The importance of the classical histopathoanatomical characteristics (e.g., tumor size, lymph node invasion, and grade of anaplasia) in predicting the course of a disease is well established. In contrast, the many varying reports of newer biochemical tissue markers generally lead to confusion. Steroid hormone receptors, the first of such markers recommended for routine determination in primary breast cancer, illustrate this point well. Since Knight et al.'s 1977 publication that estrogen receptors (ER) in tumor tissue are associated with longer recurrence-free survival (RFS) [1], there have been more than 700 published reports regarding prognosis and ER in breast cancer. However, consensus that the probability of benefit of adjuvant hormone treatment is highest among ER-positive patients was finally reached in the EBCTG study published in 1990 [2]. Meanwhile, there is no consensus regarding the value of receptor determinations for predicting the natural history of the disease. It is the purpose of this paper to address some of the more basic explanations for this uncertainty.

Some reasons for the discrepancies found in the literature are obvious. Significant differences in the size and/or profile of the patient populations, as well as differences in the observation time, lead inevitably to divergent results. Short observation times yield invalid results if there are too few "events" (i.e., recurrent disease and/or death). Apart from its importance to the validity of the results, the problem of duration of observation time manifests itself in yet another manner when the data are analyzed. While whatever factor is being measured (e.g., receptor concentration) may be an intrinsic trait of the tumor, the effect of which may persist in influencing the course of the disease, it is equally conceivable that what is being measured may represent a transient characteristic, the effect of which may be lost in time. Thus, the initial

[*]*Address for correspondence:* Susan M. Thorpe, Novo Nordisk Park, 2760-DK Måløv, Denmark.

ability of receptor status to distinguish between good and poor prognostic groups may disappear with time.

Other possible sources of differences described in literature are more subtle. Inter-laboratory assay reproducibility may be poor. While this source of variation is well-known and recognized by laboratory scientists performing the assays, clinicians who must interpret the results are not always alert to this issue. For reliable results, assay methodology must be standardized and assays remain under continual intra- and inter-laboratory control to screen for systematic errors.

Yet another subtle factor that may obscure results and prevent valid inter-center comparisons is the subgrouping of patients in preparing data for analysis. Exactly what constitutes an appropriate subgrouping of breast cancer patients? Few would disagree that patients receiving adjuvant treatment should be dealt with separately from those who do not. Here, I propose that categorization according to menopausal status constitutes an equally appropriate and biologically relevant subgrouping of breast cancer patients.

It is actually a wonder that so little attention in the field of breast cancer has previously been devoted to menopausal status. Menopause is a significant endocrine event! Endogenous levels of estradiol decrease dramatically in the menopause, and the prevalent form of available estrogen becomes estrone, which has a significantly lower affinity for binding to ER. Moreover, because ovulation ceases the corpus luteum is not formed and progesterone ceases to be produced. In considering a form of cancer assumed largely to be hormonal in nature, it would seem logical to consider the endocrine milieu of the patient at the time of diagnosis.

Epidemiologically, it is well-known that the menopause is significant. It manifests itself as an abrupt change in the age-specific incidence of breast cancer incidence, known as "Clemmesen's hook" [3]. Yet menopause is often neglected in clinical analyses. Perhaps one of the reasons for this omission is a fixation on the fact that the clinical manifestations of the disease appear to be indistinguishable in pre- and postmenopausal women. However, while the pathoanatomical characteristics of breast cancer may be similar in pre- and postmenopausal patients, examples of biochemical differences in tumor tissues are easily identifiable, provided, of course, that one looks for them.

In order to present my case for incorporation of menopausal status as an important variable in data analyses, I will use data from the Danish Breast Cancer Cooperative Group (DBCG) project [4]. The DBCG constitutes a collaborative, nationwide project initiated in 1977 that registers 95% of the primary breast cancer cases in Denmark. Thus, demographic biases often inherent in other studies (e.g., life style, social status) can be avoided by studying this patient population. In the DBCG program, patients are subdivided into treatment groups according to risk and menopausal status (pre/peri- vs. postmenopausal). Thus, information regarding menopausal status is readily available. However, it should be noted that the criteria for defining menopausal status are rather crude in the DBCG project. The DBCG observes the WHO classification criteria (≥ 5 years of menostasia constitutes menopausal status), leading to accurate identification of the postmenopausal group. Meanwhile, it remains impossible to discern between pre- and perimenopausal women on a biological basis. In our previous studies an age of 50 has been used to discriminate between pre- and perimenopausal women.

ER and PgR profiles and absolute concentrations constituted the first biochemical markers in the DBCG population, demonstrated to differ significantly according to menopausal group [5]. The frequency of ER-positivity is higher among post- than pre/perimenopausal patients, while the opposite holds true for PgR-positivity. The receptor profile of ER+PgR- is characteristic of postmenopausal women, while the profile of ER-PgR+ is characteristic of the premenopausal group. Absolute concentrations of ER increase significantly with age, while no such age-associated trends were observed for PgR concentrations.

The point that aspects of the etiology of breast cancer may be obscured by failure to perform relevant subgrouping of patients is perhaps best illustrated in our earlier work, evaluating the potential role that ER and/or PgR might play in the natural history of the disease (i.e., patients not treated with adjuvant therapy) [6]. At a median follow-up of 50 months, a complex situation was observed in 807 low risk patients. If all 807 patients were considered simultaneously, as often has been the case in the literature, a significant but small (~8% at 3 years) difference in RFS was observed with regard to PgR but not ER. However, when the patients were sub-divided according to menopausal status, much greater differences in RFS between receptor-positive and receptor-negative groups were observed for both ER and PgR (~14% and ~24%, respectively, at 3 years) in pre- but not postmenopausal patients. In the postmenopausal group the RFS curves for receptor-positive and receptor-negative patients were almost superimposed.

In a recent analysis of all patients protocolled in the DBCG 77- and 82-protocols, and with receptor determinations performed in a single laboratory, we have identified significant, clinically relevant cut-off levels for ER and PgR in each of the four biologically relevant subgroups of patients (e.g., pre- vs. postmenopausal with and without adjuvant therapy) (manuscript under preparation). Importantly, cut-off levels differ for each group. Moreover, while ER is the most important receptor for the postmenopausal women, PgR is most important for the pre/perimenopausal patient.

Since our first observation that ER and PgR are significantly associated with menopausal status, we have also found that the same holds true for other biochemical markers. The first of these was cathepsin D [7]. Not only did the distribution of cathepsin D concentrations differ among pre/peri- and postmenopausal patients, the optimal cut-off level to subdivide the populations into those with long vs. short RFS constitutes different proportions of the populations. Approximately 75% of the pre/peri- but only approximately 25% of the postmenopausal populations can expect to have longer RFS on the basis of cathepsin D concentrations.

Urokinase-type plasminogen activator (uPA) and type 1 plasminogen activator inhibitor (PAI-1), two other proteins associated with proteolytic activity, are also interesting. Multivariate analysis of established prognostic factors plus uPA and PAI-1 was recently published in a study of 190 DBCG patients. uPA, but not PAI-1, was significant in predicting overall survival among premenopausal patients. In contrast,

222

PAI-1, but not uPA, was significant for postmenopausal patients [8].

DNA ploidy, as determined by flow cytometry, provides yet another example of menopause-related differences in the significance of biochemical markers for the DBCG population [9].

Perhaps the most interesting aspect of the above differences is whether the biology of breast cancer truly differs among pre- and postmenopausal women. Only future research can answer this question. Relevant subgrouping of patients is essential if we are to extract the maximum amount of information from our data analyses in order to better understand the etiology and biology of breast cancer. It is this author's contention that menopausal status constitutes a biologically meaningful and necessary subgrouping.

References

1. Knight WA, Livingstone RB, Gregory EJ, McGuire WL. Cancer Res 1977;37: 4669–4671.
2. Early Breast Cancer Trialists' Collaborative Group (EBCTCG). In: Treatment of Early Breast Cancer. Oxford: Oxford University Press, 1990.
3. Clemmesen J. Br J Radiol 1948;21:583–590.
4. Andersen KW, Mouridsen HT, Castberg TH et al. Dan Med Bull 1981;28: 102–106.
5. Thorpe SM. Acta Oncologica 1988;27:1–19.
6. Thorpe SM, Rose C, Rasmussen BB, Mouridsen HT, Bayer T, Keiding N. Cancer Res 1987;47: 6126–6133.
7. Thorpe SM, Rochefort H, Garcia M, Freiss G, Christensen IJ, Khalaf S, Paolucci F, Pau B, Rasmussen BB, Rose C. Cancer Res 1989;49:6008–6014.
8. Grøndahl-Hansen J, Christensen IJ, Rosenquist C, Brünner N, Mouridsen HT, Danø K, Blichert–Toft M. Cancer Res 1993;53:2513–2521.
9. Balslev I, Christensen IJ, Rasmussen BB, Larsen JK, Lykkesfeldt AE, Thorpe SM, Rose C, Briand P, Mouridsen HT. Int J Cancer 1994;56:16–25.

Angiogenin is a valuable prognostic factor in node-negative breast cancer

Jörg Bläser[1], Christoph Thomssen[2], Manfred Schmitt[2], Lothar Pache[2], Fritz Jänicke[2], Henner Graeff[2] and Harald Tschesche[1*]

[1]*Universität Bielefeld, Institut für Biochemie, Universitätsstraße 25, 33615 Bielefeld; and *[2]Frauenklinik der Technischen Universität München, Ismaningerstraße 22, 81675 München, Germany*

Abstract. Angiogenesis is an important feature in tumor invasion and metastasis. Angiogenin was first isolated as an angiogenesis-inducing protein from an adenocarcinoma cell line and later found to be a constituent of normal human plasma. Angiogenin is an effective inducer of vascularization. Furthermore it has RNase-activity towards t- and r-RNA. To examine the contents of angiogenin in breast cancer tissue extracts, the newly established enzyme-linked immunosorbent assay (ELISA) procedure was used. The data showed that angiogenin has a good prognostic value in the diagnosis of disease-free survival in node-negative breast cancer.

Discovery of angiogenin

Angiogenesis is an important feature of metastasis of human tumors [1]. In 1985 Fett et al. [2] isolated the first human tumor-derived protein with angiogenic activity from the adenocarcinoma cell line HT-29. This basic single-chain protein with a molecular weight of 14400 was named angiogenin. The authors used the chick embryo chorioallantoic membrane assay to monitor the biological activity of angiogenin. The protein displays its biological activity with as little as 35 fmol per egg. Furthermore, they needed 3.5 pmol to induce blood vessel growth in the rabbit cornea. Using automated Edman degradation, Strydom et al. [3] elucidated the complete amino acid sequence of angiogenin. The protein consists of 123 amino acids. It could be concluded by peptide mapping that angiogenin contains three disulfide bonds. On the basis of these sequence data Kurachi et al. [4] were able to isolate the cDNA from a cDNA library, prepared from human liver mRNA, which they then cloned and sequenced. The gene for angiogenin was shown to be free of introns and there is only one copy of the gene in human DNA.

Angiogenin was expressed in *Escherichia coli* [5] and in cultured baby hamster kidney cells [6]. Thus, r-angiogenin became available for structural and functional studies.

Furthermore, angiogenin has RNase activity towards t- and r-RNA [7,8].

Address for correspondence: Harald Tschesche, Universität Bielefeld, Institut für Biochemie, Universitätsstraße 25, 33615 Bielefeld, Germany.

Breast cancer and angiogenesis

Tumor growth and angiogenesis are strongly linked [9]. In 1991 Weidner et al. [10] found that breast cancer patients with metastases have a significantly higher microvessel count and density grade than those who did not develop metastases. Furthermore, these data describing vascularization correlated significantly with the risk of distant metastasis. In a study with 165 patients Weidner el al. [11] found that microvessel density in the area of most intensive neovascularization is associated with poor overall and relapse-free survival in breast cancer patients. Microvessel density was only statistically significant as a predictor of overall survival among node-negative women. The authors concluded that microvessel density is a tool to select those node-negative patients who have a higher risk to relapse. They suggested that these patients should be given systemic adjuvant therapy.

Similar results were obtained by Bosari et al. [12] who examined 88 patients with axillary node-negative breast carcinoma and 32 patients with axillary node-positive breast carcinoma. They found a significantly higher microvessel count in the group of node-negative patients, who had distant recurrence. Thus, high vessel count may correlate with increased tumor angiogenesis and represent the aggressiveness of a tumor.

Angiogenin is an independent prognostic marker in node-negative breast cancer

Since the microvessel density has been developed as an independent prognostic factor in node-negative breast cancer, the question arises which angiogenesis stimulatory proteins are produced in the tumor tissue. Since angiogenin is one of these proteins, we used our newly established enzyme-linked immunosorbent assay (ELISA) method [13] to determine its content in tumor extracts. Tissue extraction was performed as previously published [14]. Other classical factors such as axillary node-status, menopausal status, hormone receptor status and tumor diameter were also included in this study. The well established factors uPA and PAI-1 [15—18] were also used in the statistical analysis. The disease-free survival was analyzed according to the Cox model [19]. The curves were calculated according to the Kaplan-Meier method [20].

From the 234 patients in our study, 97 were node-negative. The median follow up was 35 months with a range between 1 and 75 months. The median age of the patients was 55.4 years with a range from 27.3 to 88.6 years.

The univariate analysis, which covers all patients, showed a significantly higher probability to relapse for patients with an angiogenin level higher than 24.5 ng/mg protein ($p = 0.0444$) (Fig. 1). However, this difference was no longer statistically significant when a multivariate analysis was used ($p = 0.2553$) and no subdivision of node-negative and node-positive women was undertaken.

When we only examined the 97 node-negative patients we saw a very high significance in the univariate analysis ($p = 0.0087$) (Fig. 2). Again, patients with an angiogenin content higher than 24.5 ng/mg protein had an increased risk to relapse.

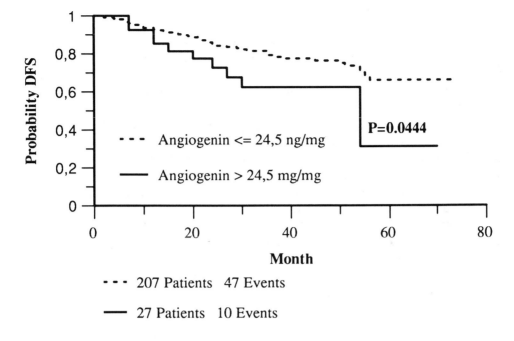

Fig. 1. Disease-free survival (DFS) as a function of angiogenin. Patients with a high angiogenin content in their tissue extracts have a higher risk to relapse (taken from the original paper: Bläser J, Thomssen C, Schmitt M, Pache L, Jänicke F, Graeff H and Tschesche H, in preparation).

In this group the multivariate analysis of disease-free survival determined angiogenin to be a very good prognostic factor. After uPA and PAI-1, angiogenin was the third highly significant predictor (p = 0.0264). The relative risk to relapse was calculated to be 4.8, with a range between 1.4 and 17.3. Angiogenin was a much stronger factor than the classical factors hormone receptor status, tumor diameter and menopausal status, none of which were significant in our group of patients.

Angiogenin correlated weakly with the menopause status (p < 0.05) and presence of tumor necrosis (p < 0.05). No significant correlation between the angiogenin level and the classical factors, axillary lymph node-status, hormone receptor status and tumor diameter, was noted.

Conclusions

Our data clearly indicates that angiogenin can function as an additional prognostic factor in the diagnosis of node-negative breast cancer in women. The determination of angiogenin with our newly developed ELISA may help to decide whether a woman should receive additional therapy or not. To our knowledge this is the first demonstration of a pathophysiological relevance of angiogenin. Our findings agree well with the data obtained by Weidner et al. [11] and by Bosari et al. [12], who showed that

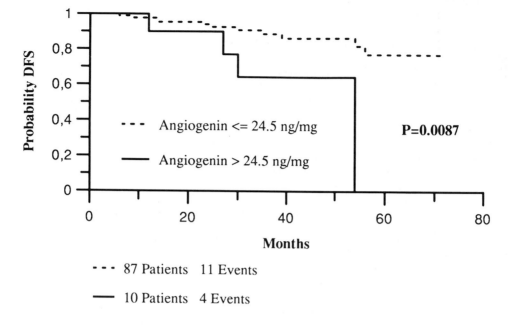

Fig. 2. Disease-free survival (DFS) in node-negative breast cancer patients. Patients with an angiogenin content of more than 24.5 ng/mg protein have a much higher risk to relapse (taken from the original paper: Bläser J, Thomssen C, Schmitt M, Pache L, Jänicke F, Graeff H and Tschesche H, in preparation).

the determination of the microvessel density has prognostic value for the disease-free survival of node-negative women. From these data the question arises whether there is a correlation between angiogenesis and angiogenin levels in the tissues. More studies are needed in this field. It should be studied, by immunohistochemical analysis, whether angiogenin is a product of the tumor cell itself or of the surrounding cells. Rybak et al. [21] found that the C-terminal angiogenin peptide Ang(108-123) decreases neovascularization induced by angiogenin in the chick chorioallantoic membrane assay. It remains to be elucidated if this peptide will have any influence on the aggressiveness of a tumor in an animal model. Since angiogenin shows good prognostic relevance in breast cancer, it should be examined whether it is a general tumor vascularization inducing agent.

Acknowledgements

This work was supported by the Deutsche Forschungsgemeinschaft, special research program SFB 223, project B1 and the Klinische Forschergruppe GR280/4-1. The authors wish to thank Mrs V. Süwer, Mrs E. Sedlaczek and Mrs B. Jaud-Münch for skilful technical assistance and Mrs G. Delany for linguistic advice.

References

1. Blood CH, Zetter BR. Biochim Biophys Acta 1990;1032:89—118.
2. Fett JW, Strydom DJ, Lobb RR, Alderman EM, Bethune JL, Riordan JF, Vallee BL. Biochemistry 1985;24:3480—3486.
3. Strydom DJ, Fett JW, Lobb RR, Alderman EM, Bethune JL, Riordan JF, Vallee BL. Biochemistry 1985;24:5486—5494.
4. Kurachi K, Davie EW, Strydom DJ, Riordan JF, Vallee BL. Biochemistry 1985;24:5494—5499.
5. Shapiro R, Harper JW, Fox EA, Jansen HW, Hein F, Uhlmann E. Anal Biochem 1988;175:450—461.
6. Kurachi K, Rybak SM, Fett JW, Shapiro R, Strydom DJ, Olson KA, Riordan JF, Davie EW, Vallee BL. Biochemistry 1988;27:6557—6562.
7. Shapiro R, Riordan JF, Vallee BL. Biochemistry 1986;25:3527—3532.
8. Lee FS, Vallee BL. Biochem Biophys Res Comm 1989; 161:121—126.
9. Folkman J. Sem Cancer Biol 1992;3:65—71
10. Weidner N, Semple JP, Welch WR, Folkman J. N Engl J Med 1991;324:1—8.
11. Weidner N, Folkman J, Pozza F, Beqvilacqua P, Allred EN, Moore DH, Meli S, Gasparini G. J Natl Cancer Inst 1992;16:1875—1887.
12. Bosari S, Lee AK, DeLellis RA, Wiley BD, Heatley GJ, Silverman ML. Hum Pathol 1992;23: 755—761.
13. Bläser J, Triebel S, Kopp C, Tschesche H. Eur J Clin Chem Clin Biochem 1993;31:513—516.
14. Jänicke F, Schmitt M, Pache L, Ulm K, Harbeck N, Höfler H, Graeff H. Breast Cancer Res Treat 1993;24:195—208.
15 Duffy M, O'Grady P, Denavey D, O'Siorain L, Fenelly JJ, Lijnen HJ. Cancer 1988;62:531.
16. Jänicke F, Schmitt M, Ulm K, Gössner W, Graeff H. Lancet 1989;II:1049.
17. Jänicke F, Schmitt M, Hafter R, Hollrieder A, Babic R, Ulm K, Gössner W, Graeff H. Fibrinolysis 1990;4:69—78.
18. Foekens JA, Schmitt M, van Putten WLJ, Peters HA, Bontenbal M, Jänicke F, Klijn JGM. Cancer Res 1992;52:6101—6105.
19. Cox DR. J R Stat Soc (B) 1972;34:187—220.
20. Kaplan EL, Meier P. J Am Stat Assos 1958;8:771—783.
21. Rybak SM, Auld DS, Clair DK, Yao QZ, Fett JW. Biochem Biophys Res Comm 1989;162:535—543.

Index of authors

Keyword index

234